AF589453

The Teaching of Chemistry and Physics in the Secondary School

MAVEN BOOKS

The Teaching of Chemistry and Physics in the Secondary School

Alexander Smith, B.Sc., Ph.D

Professor of Chemistry in Columbia University

and

Edwin H. Hall, Ph.D

Professor of Physics in Harvard University

MAVEN BOOKS

Chennai Trichy New Delhi

This edition has been published in india by arrangement with Carson Books, UK

ISBN 978-93-91270-03-2 Maven Books

All rights reserved No. 44, Nallathambi Street,
Triplicane, Chennai 600 005

MJP 1074 © Publishers, 2021

Publisher : C. Janarthanan

Publisher's Note

The legacy of a country is in its varied cultural heritage, historical literature, developments in the field of economy and science. The top nations in the world are competing in the field of science, economy and literature. This vast legacy has to be conserved and documented so that it can be bestowed to the future generation. The knowledge of this legacy is slowly getting perished in the present generation due to lack of documentation.

Keeping this in mind, the concern with retrospective acquiring of rare books has been accented recently by the burgeoning reprint industry. Maven Books is gratified to retrieve the rare collections with a view to bring back those books that were landmarks in their time.

In this effort, a series of rare books would be republished under the banner, "Maven Books". The books in the reprint series have been carefully selected for their contemporary usefulness as well as their historical importance within the intellectual. We reconstruct the book with slight enhancements made for better presentation, without affecting the contents of the original edition.

Most of the works selected for republishing covers a huge range of subjects, from history to anthropology. We believe this reprint edition will be a service to the numerous researchers and practitioners active in this fascinating field. We allow readers to experience the wonder of peering into a scholarly work of the highest order and seminal significance.

Maven Books

Editor's Note

THE present volume in the American Teachers Series follows the lines marked out in the preface to the first volume. No effort has been made to bias or harmonize the views of the authors. It has been deemed better to have a logical presentation, even at the risk of disagreement, than to give the impression that there is only one way of conceiving or giving class instruction. These volumes are intended as teachers' helps, and their purpose is served if they are suggestive to teachers who are earnestly seeking for improvement.

The authors of the separate parts on Chemistry and Physics have conferred frequently during the progress of the work, and have endeavoured to avoid unnecessary duplication. There are thus some subjects of equal importance to teachers of chemistry and physics which are discussed in one of the sections only. Some references, equally suited to either part, appear only once for the same reason. In a few instances, however, the divergence between the opinions of the authors seemed to make it desirable that each should present his own.

JAMES E. RUSSELL.

TEACHERS COLLEGE,
COLUMBIA UNIVERSITY.

Contents

THE TEACHING OF CHEMISTRY IN THE SECONDARY SCHOOL

CHAPTER I

INTRODUCTION

CHAPTER II

CHEMISTRY IN THE CURRICULUM

CHAPTER III

THE INTRODUCTION OF THE SUBJECT

CHAPTER IV

INSTRUCTION IN THE LABORATORY

CHAPTER V

INSTRUCTION IN THE CLASSROOM

CHAPTER VI

SOME CONSTITUENTS OF THE COURSE

CHAPTER VII

THE LABORATORY, EQUIPMENT AND ILLUSTRATIVE MATERIAL

CHAPTER VIII

THE TEACHER, HIS PREPARATION AND DEVELOPMENT

THE TEACHING OF PHYSICS IN THE SECONDARY SCHOOL

CHAPTER I

WHETHER TO BE A TEACHER OF PHYSICS

CHAPTER II

PREPARATION FOR TEACHING

CHAPTER VII

THE TECHNIQUE OF LABORATORY MANAGEMENT

CHAPTER VIII

LECTURES AND RECITATIONS

CHAPTER IX

PHYSICS IN PRIMARY AND GRAMMAR SCHOOLS

CHAPTER X

PHYSICS IN VARIOUS KINDS OF SECONDARY SCHOOLS

Prefatory Note

WHILE there is only one science of chemistry, there are many opinions on the teaching of it. A tendency to seeming dogmatism is almost unavoidable in writing on the subject. It can only be said that, by an attempt to present the various views, even where there is a very distinct inclination towards one view on the part of the majority of chemists, an effort has been made to avoid real intolerance. The references to articles and books dealing with all controversial points will enable the reader to acquaint himself with the best that can be said for each opinion by its special advocates and to reach a decision for himself. Where the writer has felt certain that fundamental faults are committed by any considerable number of the less well-trained teachers, as, for example, in connection with equations, the atomic theory, natural law, etc., the view held to be correct may appear to have been presented in an extreme form or with needless vigour. Perhaps it is hardly necessary to explain that the precise treatment given here in such cases is not necessarily suited as it stands for the consumption of the pupil. If it impresses the teacher strongly by putting a different view in high relief, and shows him the standpoint of the chemist as opposed to that of some pedagogues, it will have served its purpose.

The book assumes the reader's familiarity with the science. Many brief references to chemical matters would have been greatly expanded or carefully limited if they had been intended for beginners.

There has been no intention to recommend particular text-books. Many have been cited as possessing some point of special excellence, but this must not be taken to indicate emphatic approval of those works as wholes, say for use with a class of beginners.

The author is indebted to many friends for helpful suggestions. Special thanks are due to Messers. J. B. Tingle, Illinois College, M. S. Walker, West Division High School, Chicago, W. A. Noyes, Rose Polytechnic Institute, and J. B. Garner, Wabash College, who have read all the work in manuscript or proof.

ALEXANDER SMITH.

THE UNIVERSITY OF CHICAGO,
March, 1902.

The

Teaching of Chemistry in the Secondary School

CHAPTER I

INTRODUCTION

BIBLIOGRAPHY.

Spencer, Herbert. Education, Intellectual, Moral and Physical. London, Williams & Norgate. New York, D. Appleton & Co. 1860.

Bain, Alexander. Education as a Science. London, Kegan Paul, Trench & Co. New York, D. Appleton & Co. 1881.

Macgregor, J. G. The Utility of Knowledge-making as a Means of Liberal Training. NATURE, LXI. (1899), 159–163; extracts in SCHOOL REVIEW, VIII. (1900), 372.

Eliot, C. W. Educational Reform. New York, The Century Co. 1898.

Mach, Ernst. On Instruction in the Classics and the Sciences. Popular Scientific Lectures. Chicago, The Open Court Publishing Co. London, Kegan Paul, Trench & Co. 1895.

Coulter, J. M. The Mission of Science in Education. SCIENCE [N. S.], XII. (1900), 281–293.

Wilson, C. C. What is the Consensus of Opinion as to the Place of Science in the Preparatory Schools? SCHOOL REVIEW, VI. (1898), 203–211.

Huxley, T. H. Science and Education. Collected Essays, Vol. III. London and New York, Macmillan. 1893.

Perkin, W. H., Jr. The Modern System of Teaching Practical Inorganic Chemistry and its Development. Vice-Presidential Address. Report of the British Association, 1900. Reprinted in NATURE, LXII. 476–481.

Perkins, A. S. Elementary Chemistry in the High School. SCHOOL SCIENCE, I. 72–77.

Williams, R. P. The Teaching of Chemistry in Schools, 1876–1901. SCIENCE [N. S.], XIV. 100–104.

Long, J. H. Early History and Present Condition of the Teaching of Chemistry in the Medical Schools of the United States. Vice-Presidential Address before the American Association for the Advancement of Science (1901). Reprinted in SCIENCE [N. S.], XIV. 360–372.

Russell, J. E. German Higher Schools. New York and London, Longmans, Green & Co. 1899. Pp. 329–351.

Cooley, LeR. C. Science for Education. Presidential Address before the New York State Science Teachers' Association. High School Bulletin No. 7. Albany, N. Y., The University of the State of New York. Pp. 578–594.

Butler, N. M. The Scope and Functions of Secondary Education. EDUCATIONAL REVIEW, XVI. (1898), 15–27.

IN introducing the subject of the teaching of chemistry it is fitting first to state as briefly as may be some of the reasons for the inclusion of this study in the curriculum of the secondary school. We shall thus have in hand the key to the point of view which will be preserved in the treatment of the subject as a whole, and our statements will indicate the ideal towards which the means of education that we may discuss are to be directed. Since the ideal which we set before ourselves must be brought into relation with the actual conditions, it will be advisable also to say something in regard to the present state of chemistry in the secondary schools of this country. We shall then be able more easily to determine the means by which the actual may be converted into, or at least be brought to approach this ideal.

Spencer's Reasons for the Study of Science.

In advocating the study of a science, all intention of regarding it as the only worthy means of education, and of urging that it shall supplant other subjects, must be discarded. It is true that the support of no mean authority might be found even for this radical position. Herbert Spencer, in his chapter on "What knowledge is of most worth," attempts to show, and with some success, that the information which the study of science furnishes is incomparably more useful for our guidance in life than any other kind. Starting from the thesis that education is intended to teach us how to live, and considering all the activities of human life, he demonstrates that science furnishes equally in all cases the needful preparation. Considering next its use as a discipline, he concludes "that for discipline, as well as for guidance, science is of chiefest value. In all its effects, learning the meanings of things, is better than learning the meanings of words. Whether for intellectual, moral or religious training, the study of surrounding phenomena is immensely superior to the study of grammars and lexicons." He contends that the value of the information furnished by history and consisting of a mere tissue of names and dates has a conventional value only. On the other hand an acquaintance with Latin and Greek, since it furnishes extra knowledge of our own language, but can be of value only so long as certain languages and races exist, may be held to have a value that is at best quasi-intrinsic. The knowledge furnished by the truths of science, such as that resistance to the motion of a body varies as the square of the velocity, and that chlorine is a disinfectant, are truths which will bear on human conduct ten thousand years hence as they do now, and can therefore alone lay claim to true intrinsic value.

While we cannot doubt the great and in many ways entirely wholesome influence which Spencer's views have exercised, we may be permitted to point out, that, in supporting the claims of some particular subject or class of subjects to inclusion in a course of education, one is always in danger of comparing the subject he favours *at its best*, and in the ideal form it may receive at the hands of an ideal teacher, with other subjects at their *average* or even at their *worst*. This essay was written before 1860, and was published in that year with the others forming the volume on education. Perhaps the circumstances of that time justified much of what is said, in a way that the circumstances of the present would not. There have undoubtedly been great improvements in language teaching at its best since that time, and while Spencer states that "science forms scarcely an appreciable element in our so-called civilized training," and in another place speaks of it as "that which our school course left almost entirely out," these statements could not be made with justice in the same form at the present time.

Our task will be rather to show that science undoubtedly has an aptitude for educational employment sufficient to make it a valuable study. We may fairly say also that its use is justified by the desirability of introducing variety into our means of instruction. This must increase its strength and effectiveness in view of the many-sidedness of the interests and therefore of the avenues of approach in each individual case, and of the differences in the tastes of different individuals. We may perhaps even go so far as to demand its employment on the ground that in some directions, even at its average, it can furnish certain constituents of an all round discipline of the mind more easily or in a more conspicuous degree than other subjects at their average.

Other Reasons for the Study of Science.

We must begin with characteristics which are common to all science, with physical science chiefly in mind, and afterwards more briefly refer to the special claims of chemistry. The order in which the reasons for the study of science are given is not so much that of importance as that of convenience of presentation.

I. Reasons for the Study of Science.

Training in Observation.

Our first reason for the study of science rests on the training in observation for which it furnishes the opportunity. To use the common expression, the employment of the laboratory method in science furnishes an exercise not merely for the senses but for the mind. This practice of observation is, however, common to all forms of scholarship, and is applied in languages, for example, as persistently as in science. But in the latter this training directs attention to material objects, and so, while theoretically the same process, it awakens an interest in, and develops a capacity to exercise an entirely different and most worthy set of activities. Its application to the study of nature and her laws gives, as nothing else can, a distinct view of the universe as a well-ordered system. As President Eliot says (*Educational Reform*, 110), "The student of natural science scrutinizes, touches, weighs, measures, analyzes, dissects, and watches things. By these exercises his powers of observation and judgment are trained, and he acquires the precious habit of observing the appearances, transformations, and processes of nature. Like the hunter and the artist, he has open eyes and an educated judgment in seeing. He is at home in some large tract of nature's domain." Mach (*Scientific Lectures*, 280) makes the same point as follows: "I shall meet with no contradiction when I say that without at least an elementary mathematical and scientific education a man remains a total stranger in the world in which he lives, a stranger in the civilization of the time which bears him. Whatever he meets in nature, or in the industrial world, either does not appeal to him at all, from his having neither eye nor ear for it, or it speaks to him in a totally unintelligible language." This interest in a knowledge of nature is an essential element in robust life. The study of books alone, at its worst, often submerges the victim of it in an unpractical and even mediæval spirit which we all recognise as characteristic of the bookworm.

Training in Comparison and Induction.

The second reason for the study of science is that it trains the pupil in the organization of his observations by comparison and induction. This again is an operation not by any means peculiar to scientific work. Every human being from the earliest months of his existence onwards is occupied in the co-ordination of the observations he makes, and in building up, with rapidly increasing speed, a mass of rationalized experience. The business of life, when it is entered upon, is promoted by the same processes. Consciously or unconsciously observations are made and generalizations produced from them, and on the ability to do this correctly the man depends for his success in life. This has been called by Professor J. G. Macgregor[1] the "knowledge-making" process, by which he means that separate facts do not constitute knowledge in themselves, and that all we know which is of value is *made* by putting together our isolated observations. It is our ability to do this which counts in scholarship or in life, and it is this therefore which education should be specially directed to cultivate. As Professor Macgregor points outs, the old curriculum, and particularly the study of the classics, affords abundant op-

[1] In a most interesting inaugural address delivered at the opening of the fortieth session of Dalhousie College and printed in full in NATURE.

portunity for the exercise of this power. The study of language involves the continual putting together of instances of the use of words and phrases. "The lexicon," he continues, "would give little more than a clew in many cases to the English equivalents of, say, Latin words, the exact equivalents, whether words or phrases, being determinable only by the study of the context and the fruitful drawing upon experience."

Differences between Work in Languages and in Science.

There are two differences between this process as carried out in the study of a language and the same process employed in the study of a science. In handling the grammar, dictionary and text, it is almost impossible for this instruction in its early stages, and at all events at its average if not at its best, to avoid confirming the tendency usually fostered by ordinary home training of relying upon authority for information and opinions. Even the most elementary work in science must be considerably below the average if it does not place the experimental facts themselves before the pupil, and suggest to him no other ultimate source of information, whether for the learner, the teacher, or the author of the text-book, than the study of nature itself. The other advantage which natural science possesses is that experiment permits the repeated production of every fact fresh from its original source in the order of nature, as often as may be necessary for its full appreciation, and with such variations in its circumstances and surroundings as a clear view of every side of it may demand. The method of experiment is a tower of strength in the study of the facts of science, while its application to languages is usually discouraged by the teacher, and its use in history is impossible. The recourse in this way to the study of the thing itself for first hand information, the exactness and conclusiveness with which every fact may be established, the testing of every inference or hypothesis by renewed comparison with facts, form the essence of the scientific method. While it is employed as far as possible in all scholarly work, yet it was first recognised by logicians in connection with the rapid growth of science which resulted from its systematic application in the study of nature, and it still finds special opportunities of employment along scientific lines which are lacking in other directions. It is not too much to expect that the practice of this universally applicable method on the original material will form an invaluable foundation for its subsequent use in any direction whatever. As Professor Jebb says: "The diffusion of that which is specially named science has at the same time spread abroad the only spirit in which any kind of knowledge can be prosecuted to a result of lasting intellectual value."

Use of the Imagination in Science.

Amongst the possibilities which the study of a science prominently presents is that of the exercise and control of the imagination. In connection with every subject of thought it is a function of the mind to rearrange the various conceptions which are presented, to recombine them in new forms, and to invent hypotheses which justify the new combinations. Of all the powers of the mind it is certainly that which is most important in giving originality to the results of thought. But in proportion to its value and activity is the difficulty in controlling its operations. The imagination is a good servant, but a bad master. The opportunity which is offered in an experimental science to test the results of the work of the imagination by comparison, again and again renewed, with the concrete materials with which it has been dealing, furnishes an unrivalled opportunity to practise the control of it. To parody Dr. Johnson's aphorism about Richardson, the study of science must do much towards teaching the imagination to move at the command of truth.

As Professor John M. Tyler says[1]: "The successful scientist will always exercise his own imagination and that of his pupils. He will not allow 'This valuable gift of nature to be repressed by a bookish and wordy education.' He will encourage no day-dreaming fancy. He will demand that the pictures of the imagination shall be rigidly tested to see that they correspond to

[1] The Culture of the Imagination in the Study of Science; SCHOOL REVIEW, VI. (1898), 716. The whole article should be read. See also Pearson's *Grammar of Science*, chapter I.

some objective reality. But within these limits, and with these restrictions, the student of science will cultivate his imagination as faithfully as the student of art. And he will train it and control it with far more scrupulous fidelity."

Self-elimination in Science Work.

The study of science gives training in what, for want of a better name, we may designate self-elimination. In all subjects clear and unbiased judgment is the ultimate goal of the student's effort, but many branches of knowledge are so filled with human opinions, permeated by conventional standards, and all through show so strongly the stamp of human workmanship, that unprejudiced judgment is hard to attain. Professor Coulter, in an address delivered at the University of Michigan, presents clearly the idea I wish to convey: —

"The general effect of the humanities in a scheme of education may be summed up in the single word *appreciation*. They seek so to relate the student to what has been said or done by mankind, that his critical sense may be developed, and that he may recognise what is best in human thought and action. To recognise what is best involves a standard of comparison. In most cases this standard is derived and conventional; in no case is it founded in the essential nature of things, in absolute truth, for it is apt to shift. In any case the student injects himself into the subject; and the amount he gets out of it is measured by the amount of himself he puts into it. It is the artistic, the æsthetic, which predominates, not the absolute. It is all comparative rather than actual. The ability to read between the lines is certainly the injection of self into the subject-matter, and the whole process may be regarded as one of *self-injection* in order to reach the power of *appreciation*. . . . The proper and distinctive intellectual result of the sciences is a *formula*, to obtain which there must be rigid *self-elimination*. Any injection of self into a scientific synthesis vitiates the result. The standard is not a variable, an artificial one, developed from the varying tastes of man, but absolute, founded upon eternal truth."

Of course neither of these qualities of self-injection and self-elimination is confined to the branch of knowledge in which it is thus discovered to be a conspicuous constituent. As Professor Coulter points out, even in the study of science alone the self-injecting tendency of the humanities must be combined with the self-eliminating tendency of science, "and the power of appreciation developed by the humanities must always be tempered by the scientific spirit." Each method is needed in the study of both. Objective and dispassionate study is the result of the application of the scientific method in any field of knowledge, but the material of the sciences favours the achievement of this ideal in an especial degree. As Huxley says, the scientific mind is "a clear, cold, logic engine, with all its parts of equal strength, and in smooth working order."

Other characteristics of mental discipline, such as the way in which it stimulates mental rectitude, favours clear thought, leads to clear expression, etc., might be discussed. They are, however, either implied more or less distinctly in those already enumerated, or may be discussed more fitly in connection with the particular parts of chemical work which call out their employment.

Value of the Information furnished by Science.

An argument of a different kind but of no small weight is founded upon the value of the information which the study of science imparts. It is at once evident that a study which has strong claims on account of its disciplinary value must become practically indispensable if it is able simultaneously to furnish information which is useful and can be obtained in no other way. It is one of the strong points of Spencer's argument in the chapter on "What knowledge is of most worth" that the information furnished by science is of this kind. In learning how to live we must consider the activities which make up life. Spencer divides these into those concerned with self-preservation, those concerned with the gaining of a livelihood, those concerned with the rearing of offspring, those which minister to the regulation of conduct in social and political relations, and those which minister to the gratification of our tastes and feelings. He shows in great detail how all-important scientific knowledge is for success in any of these lines. Huxley has used the same argument, and puts it in a very striking form. The memorable passage in one of his essays,[1] in which, after comparing life to a momentous game with complicated rules, he asks whether a parent who failed to teach his son the rules of this game would not be considered negligent or even cruel, is so familiar that to recall it is almost superfluous. In his simile, life is like a game of chess with an invisible opponent, who may be regarded if you will as some calm, strong angel, who plays for love, as the saying is, and would rather lose than win. The rules of the game are the laws of the universe. The incompetent player is checkmated — without haste, but without remorse.

Not only is this information valuable, nay, indispensable, in itself, but it assists in holding the interest of the majority of pupils who would not be so much attracted by a purely disciplinary study. Contact with physical science when it is presented in the right way must continually be felt to be contact with a mighty, living, and growing reality.

There is still another consideration which cannot be left out of account when the introduction of a subject into the curriculum is proposed, and that is that we have no reason to believe on historical grounds that the old curriculum has become so much a part of the development of the race that any change in it would produce serious organic disturbance. As President Eliot has clearly shown,[2] the classical curriculum is only three hundred years old, and displaced after a severe struggle an entirely different course of training, consisting of scholastic theology and metaphysics, while this in turn was simply the usurper of a throne formerly occupied by grammar, rhetoric and logic, with a little mathematics, music and astronomy. Thus a complete revolution of the whole list of studies might occur again without scholarship

History of the Curriculum.

[1] A Liberal Education, and where to find it. *Essays*, Vol. III.

[2] What is a Liberal Education. *Educational Reform*, 94.

becoming a byword or education perishing from the earth. Even if science and modern languages entirely displaced the classics, we could not urge that a break in the course of nature had occurred, but only that history was repeating itself. This displacement, however, is not demanded. All we make is a plea for the fair representation of science, and this must be regarded as conservative, whether we look at it in the light of reason or of history.

Summary.

To sum up, we may fairly claim that when science is employed in a way which constitutes an approach to the realization of its possibilities, it furnishes a field for observation along a special line, that of the phenomena of nature; it exercises us in knowledge-making, and for this furnishes a method of unusual power, that of the study of concrete objects and of experiment; it gives employment for the imagination, and at the same time provides an especially sure means of controlling its operations; it trains the judgment by the way in which the nature of its subject-matter favours self-elimination; and finally, the information which it yields is of a special and particularly valuable description. Other subjects may claim to provide discipline under every one of these heads, but in each case science gives a particular variety of this discipline which is distinctive. It is thus an indispensable complement to other branches of study, and, it may be added, is indispensable not merely in the secondary school, but at every stage of education.

This is the least that can be said. Many scientific men claim more for it. In ascending the Brocken one sees, at intervals of a few hundred feet, portions of ground fenced off and used for the rearing of small trees which are afterwards to be planted out at the same altitude.[1] If the same trees were sprouted at the experiment station at Goettingen, and then planted at the top of the Brocken, so far as even an authority on forestry could see, they might appear to be perfectly

[1] I have adapted this illustration from the highly suggestive presidential address of Professor Dennis on "The School and the World" delivered before the Indiana State Science Teachers' Association (1898).

adapted to the purpose, yet they would have less chance of surviving in their changed environment. It is to be feared that a large part of what we learn in school does not survive transplantation to the climate of the world, in spite of the demonstrable value of its educational possibilities. Perhaps before we have concluded our study of the teaching of chemistry, it may appear that science at its best comes nearer to being a form of training sprouted at the right altitude, and stands a better chance of thriving in every-day life than book learning in any other branch.

Some objections have been urged against the conspicuous employment of the sciences as aids in education. It is said that the method of science is so rigorous and exact that it unfits men for dealing with human questions which have not the same clean-cut qualities. Perhaps the best answer to this is that experience has not shown that men, even when they have devoted their whole lives to the study of science, very generally lose the qualities of sympathy and humanity. And, if there is no great evidence of this in their case (and, in this connection, surely the quotation of Darwin's experience has been decidedly overworked, and been manipulated like a stage army until it has furnished an apparent basis for generalization), there is little danger of serious atrophy of the sensibilities when scientific work of an elementary character occupies but a fraction of the time of our youth.

Objections to the Study of the Sciences.

It is said also that the study of science tends to lower the ideals of the student, since it calls upon him to soil his hands by contact with that which is commercially useful, and that its general pursuit would convert us into a money-getting and ease-loving people. I believe that the writing and publication of books for beginners in various languages is a very lucrative occupation, but I have not heard that this fact diminishes the cultural value either of the instruction given by the author or of the study of his books. Seriously, it is not proposed that a course should be given which shall in the remotest way suggest how one may become wealthy by the employment of chemistry. There are many teachers of chemistry who would be delighted to attend such a course, however, if it were given.

Finally, it is stated that the scientific man exercises himself with so limited a part of human experience, as compared with that touched by the classics, for example, that the study of science has of necessity a distinctly narrowing influence. This argument, and the preceding one as well, simply seem to be fresh cases of comparing one study at its worst with another study at its best. Science at its worst under a poor teacher is doubtless as narrowing, not to say dull and useless, as any other study taught under the same conditions.

The Sciences in the Curriculum.

So far we have spoken of science in general with the thought of physical science particularly in our minds. Some of the arguments in its favour hold with special force in regard to the biological sciences, and indeed, if these sciences had been considered, at least one weighty addition might have been made to the list. We are not here concerned with sciences other than physical; nor does any question arise as to whether physics is to be preferred to chemistry or chemistry to physics. Chemistry can be studied only through physics. The latter science is more easily approached, furnishes a broader field and more points of contact with every-day life, and is indeed a prerequisite to the study of any science. If chemistry is studied before, or instead of physics, about half the time at the disposal of the teacher of chemistry must be devoted to the study of physics, or the work in chemistry itself will be trivial and superficial.

Chemistry shares with physics all the characteristics of scientific study which we have discussed. It differs from it slightly, however, in respect to the degree and manner in which some of them are represented. Thus in chemistry observation is mainly through the study of physical phenomena; chemical observation is therefore made by inference and not directly. Again, the facts of chemistry which have to be taken into consideration, even in elementary work, are much more numerous than is the case in physics. The memory is much more heavily taxed in their mastery, and the powers of organization are called more prominently into play in their arrangement. Without unusual emphasis on organization and arrangement the science disintegrates into a mass of details. Finally, the study of chemistry, on account of the indirect method of observation, gives more employment to the imagination than does physics. The theories and hypotheses of chemistry are more indispensable to the appreciation of its facts. Of course, as regards information, the sciences are all distinct from one another, and there is little to choose between physics and chemistry in the matter of the importance of this.

II. History of the Teaching of Chemistry.

While the discussion centering round arguments like those we have set forth was going on, the sciences had been slowly ripening on the didactic side against the time when they should gain free representation in the schools as disciplinary studies.

The course of laboratory instruction in the earlier days of all institutions of whatever kind seems to have been modelled on that pursued by Liebig in Giessen. Here, Professor Perkin says, "after preparing the more important gases," the student "was carefully trained in qualitative and quantitative analysis." The publication of an outline of Liebig's course in qualitative analysis by Professor Will in 1846 (this was the first systematic introduction to the subject for the use of beginners), and its translation into English, seem to have been followed by the adoption of this subject as almost the sole material of elementary laboratory instruction. Even at the present day the tradition still retains its influence, and the importance of fuller preliminary instruction in general chemistry is still fighting its way to recognition in some quarters. The one-sided and distorted view of chemistry which the standpoint of elementary qualitative analysis gives is still unfortunately the only one offered to many beginners.

Methods of Instruction in Chemistry.

The first really decisive step in the right direction was taken by Harvard College, when, in 1888, it included chemistry for the first time among the subjects that might be offered for admission, and, through Professor Cooke's *Laboratory Practice*, issued in the same year, defined the kind of work which it considered to be the true educational equivalent of the other and older preparatory school studies. This book practically launched a new ideal in chemical instruction. It was the first attempt in this country[1] to lay out a course of laboratory work dealing adequately with the fundamental facts and laws of chemistry. The result was in marked contrast to the miscellaneous experimentation with chemical substances, and the dabbling in qualitative analysis which had hitherto been in use and had afforded so little support to the classroom work on the principles of the science.

In Great Britain the question of the best methods of teaching elementary chemistry had been under discussion for some years before this time. In 1889 a committee of the British Association, appointed to consider the subject, submitted a report which included a detailed outline of work (prepared by Professor Armstrong for the Committee) that was a vast improvement on the old schemes commonly in use. It resembled the American plan in discarding analysis, but was intended for a younger set of pupils, and differed from it besides in placing emphasis even more strongly on the *method* of instruction at the expense of the completeness of the account of the science. It sought to place the pupil as completely as possible in the attitude of a discoverer, and was willing to sacrifice much in the way of speed and area covered to accomplish this. In their different ways both were notable contributions to the didactic side of chemistry, and we may trace directly to their influence the rapid spread of more rational methods of teaching the science which has been so conspicuous in both countries. The principle

---

[1] Professor Ramsay's *Experimental Proofs of Chemical Theory for Beginners* (London, Macmillan, 1884) was the corresponding "first" in Great Britain.

underlying Professor Cooke's plan has been adopted by nearly every recent work prepared for the use of secondary schools. The *motif* of Professor Armstrong's report has coloured, where it has not controlled, a large proportion of the recent teaching of Great Britain.

Chemistry in the Schools of Germany.

In Germany some work in science is given in every year of the secondary school course. The instruction, however, seems to be undertaken with a different aim from that kept in view in the English-speaking countries. While in this country chemistry is taught as a separate department of instruction, and with the object of conferring some knowledge of the science itself, in addition to the extraction of such general mental discipline as its study is able to afford, in Germany the science is employed primarily for the purpose of giving intelligent knowledge of things around us, and as a means of training the powers of observation, reasoning, and exact expression. While this is the general tendency, however, there are many schools in which individual laboratory work by the pupils is included in the curriculum, and in which the place of chemistry is more like that which it holds in America to-day. Yet even in these schools the laboratory work does not call for fresh study of new problems in an independent spirit, but only repetition of those already used and illustrated in the classroom. Then, too, attendance on the laboratory exercises is optional and is usually very limited. Promotion to the next class is in theory dependent on proficiency in scientific as in other studies, but in practice takes place irrespective of this. Thus, in the German school, science, to use the words of Spencer, is "the household drudge, who, in obscurity, hides unrecognised perfections." At least these perfections, if recognised, are veiled until the regular study of the sciences as separate subjects begins in the university.[1]

---

[1] See James E. Russell, *German Higher Schools*, 329–351, in which a detailed outline of the work in physical science in one school is given, and the methods and ideals are fully set forth.

III. The Present Condition of Chemical Instruction.

The history of secondary education in America so far as chemistry is concerned, has been marked by two conflicts, first, the struggle for admission, and then, the struggle for rank. The struggle for admission may be said to have been completely won, although there may be outlying portions of the field which have not yet been occupied by the victor. The number of pupils taking chemistry in all the secondary schools in the United States is some indication of this. According to the report of the Bureau of Education, 8.55 per cent of all the pupils were studying chemistry in 1897–98. When we consider that chemistry is very rarely taught during more than one year, and that it is usually placed in one of the later years of the course, it is probable that not more than 15 per cent of all the pupils during any one year have any opportunity to take chemistry. The actual number, therefore, is, on the whole, encouraging. Attention may be called in passing to the peculiarities, either in the opportunities for taking chemistry, or in the way in which advantage is taken of these opportunities, or both, in different parts of the country. The percentages are 9.39 in the North Atlantic States, 12.22 in the Western States, 7.58 in the North Central States.

The Struggle for Admission.

The struggle for rank may be said to have been won also, but by a moral victory. The opponents are defeated, but it may be doubted whether they are convinced. They cover their retreat by the statement that the scientific course is possibly of equal value with the classical, but the training which it gives is different. On this ground they frequently would refuse the granting of the degree of A.B. to the graduates of such a course. Unquestionably the sciences have been at a disadvantage on account of their lack of that prestige which three hundred years of continuous employment have given to the older subjects. But, putting aside all prejudice, there is perhaps still some ground for reserve in answering the question whether science has actually fulfilled its promises.

The Struggle for Rank.

That its present average efficiency is far below its possible best, no one can doubt. So far as the feeling of which we have spoken is due to a distrust of science at its best, the question has already been disposed of in the first section of this chapter. So far as it expresses a lack of confidence in science, and particularly in chemistry as it is, some further inquiry into the question will be proper in this place, and before we pass to the discussion of the means which in many schools at least have enabled it to reach the best.

Difficulties with which Chemistry Contends.

There are unquestionably some things which diminish the effectiveness of chemistry as a means of instruction in our schools. The first of these is a lack of organized instruction in scientific matters running through every year of school work, from the first to the last. The pupils have not been brought up to the study of nature and physical science by personal handling of the objects with which these deal, and consequently their ability to get the best out of the subject is hampered because their capacity to employ the means of study has become partially atrophied by disuse.

This state of affairs is certainly being remedied, but that the improved conditions have yet had time to affect the average teaching in chemistry, may be doubted. Contrast this with the custom of continuously employing the methods of language study from the earliest years, which every child has acquired, and the disadvantage under which chemistry labours is at once apparent.

Defects in the Means afforded for Training of Teachers.

Still another impediment may be found in the instruction in chemistry in the higher institutions which are intrusted with the duty of training the teachers. As before, we are speaking of the average and not of the best. It has been asserted, and with justice, that the greater part of this training is essentially non-scientific in its tendency. The instruction is too dogmatic, and books are too largely the reliance of the instructors. The pupils are not disciplined in the methods of observation and investigation, and there is too much speculation substituted for the much more conservative theorizing and explanation which are alone permissible. The pupils are not brought in contact with the spirit of the subject by the study of the original sources and the memoirs of investigators, and, above all, they are scarcely ever called upon to perform any original investigation, no matter how simple, on their own account. Such instruction can never transfuse into the minds of the pupils any notion of the spirit of the subject. Take, for example, the conventional course in chemistry. Even if the subject is studied for two or three years, which occurs in the preparation of a small minority only of the prospective teachers in secondary schools, the time is largely occupied with quantitative and qualitative analysis, subjects which should play a subordinate part in the preparation of the teacher, unless he has ample time at his disposal. It is general chemistry that he must know, and instruction in these other branches not only contributes little to his knowledge of the main trunk, but even diverts his attention from it. Comparison with the training given in languages and mathematics shows that, although it may be easy to point out defects, the preparation is, after all, much more thorough in the directions in which the work of teaching in the secondary school makes the heaviest demands.

The results of faulty training of the teacher are more serious in science than in language. As Professor Macgregor says, "In the making of linguistic knowledge, a pupil under an incompetent teacher does not stick fast. He has the experience of his childhood to help him, is capable of exercising the knowledge-making power without the teacher's aid, on the familiar material which language affords, and in his effort to make progress, cannot help exercising it to a greater or less extent. Let me draw special attention to this point; for the fact that in the study of language, exercise of the knowledge-making power is not only possible, but in a large measure unavoidable, even under an incompetent teacher, gives to language study a great advantage over science study, as a means of discipline in all educational institutions, but especially in those of lower grade, in which, owing to their large number, the difficulty of securing competent teachers is especially great."

Lack of Unity in Aim and Method.

Still another cause of diminished effectiveness in chemistry teaching is the lack of unity in the aims and methods of the teachers. This is the result of the existence of the same fault in the work of the higher institutions. Some elementary courses in chemistry are devoted largely to analysis. In others, the discourse is mainly of atoms. These, instead of being employed as conceptions rather than facts, are described with such realism that the study of the subject by experiment is pressed into the background, either actually or in the estimation of the pupils. This lack of unity is so notorious that when, a few years ago, a set of educational conferences was called at Columbia University, no conference on science was held. It was considered that the opinions of its advocates were so unsettled that the colleges had no basis on which to fix definite requirements in science at all. We have but to look at text-books on chemistry in order to see that, although they are all labelled chemistry, their content and spirit differ widely. We have but to compare them with the standard treatises on languages and mathematics to see how much greater the unity is which has been reached in these subjects.

The Intrinsic Difficulty of the Science.

In enumerating the disadvantages under which the teacher labours in fitting chemistry for a place in the curriculum, we must note that not the least of these is the difficulty of the subject itself. To quote Professor Macgregor again, "A difficulty with which the sound teaching of science has met, arises from the complex character of its subject-matter. To compare different usages of words, for example, one has but to turn over the leaves of a book; to compare instances of the occurrence of natural phenomena, the phenomena must be watched for or reproduced under varying conditions." Or again, as Professor Cooley says, "Phenomena are the symbols in which truths are written, but phenomena abound in superficial likenesses, obscure differences and deceptive analogies. A correct translation of this language requires keen perception, accurate judgment and crystalline forms of expression." It is undoubtedly harder to carry the subject to a depth corresponding to that which would be reached in French or Latin, or to master it with equal thoroughness.

Trivial Nature of much School Work in Chemistry.

The case is often made worse by putting chemistry before physics in one of the earlier years in the secondary school. The highest benefits can be got from its study only when the time comes at which, as Professor Nicholas Murray Butler, in attempting to define the stages of psychological development and ascertain their correspondences with the stages in our educational system, says, the soul "demands new and more difficult problems to occupy it and absorb its activities." As we hope to show later, the organization of the teaching of chemistry at its average is in need of very great improvement before adequate benefit can be conferred by it, even in the fourth year of the secondary school. At present much time is wasted on the study of superficial aspects of this science, when the same time devoted to languages or mathematics might have gone much deeper and been educationally much more effective. Much testimony is available to prove that the chemistry work in the average school is not a trial worthy of the powers of the pupils. To use the expressive phrase of a friend of mine, "There is too much chicken-feed chemistry occupying time that might have been devoted to the giving of solid nourishment." One needs but to visit a number of schools to see that there is truth in this statement. I have seen work in English being done by the freshmen of a high school which showed a surprising grasp of the more abstract aspects of rhetoric and an ability to handle problems of literature in a wonderfully effective manner, while the same pupils in the next year were puttering with a kind of chemistry which, it may be said without exaggeration, would not have over-taxed the ability of a reasonably intelligent infant if its physical development had permitted it to attempt the work. A mode of study in a science which does not take full advantage of the knowledge-making power which it can call forth, not only largely wastes the time it occupies and discredits the science itself, but diminishes the efficiency of the whole curriculum of instruction. There is reason to fear that chemistry has gained admission before the means of using it most effectively have become widely known.

Some Other Hindrances.

The work in science is also frequently hampered by the attitude of the authorities of the school, who may not be as fully convinced of its value as their introduction of the subject into the curriculum would lead us to expect. They are apt to promote pupils who have neglected scientific work, provided they have done well in other studies. They are apt also to appease the clamour for representation of science in their school by assigning classes in chemistry to teachers who have had almost no preparation in the subject, instead of delaying its introduction until they can afford to obtain a properly prepared instructor. They are prone to load four or five sciences on one teacher, regardless of the utter impossibility of organizing good laboratory instruction under such circumstances, even if the preparation of the teacher should, by a miracle, be not unequal to the task. They cut the day into equal and often very brief periods, as if mechanical adjustment of time were everything, and the essential differences between laboratory work and class work, in respect to the value which each can get out of thirty or forty minutes, were nothing.

Summary.

To secure instruction in science of effectiveness equal to that in other subjects, and to wrest from it the benefits which it admittedly can confer, we must have continuous instruction in science, beginning with nature study in the elementary schools; we must have at the other end improvement in the chemical curricula in the highest institutions which furnish the teachers; we must have unanimity, or some approach to it, in regard to the aims and methods of secondary school chemistry; and we must work out the detailed organization of the teaching of chemistry more fully. When these things have been accomplished, proper respect for the subject at the hands of all educational authorities will come of itself. At present the average instruction in chemistry does not even remotely approach, in the benefits which it gives, the best that can be given or is given. When the difficulties we have enumerated have been removed, or considerably reduced, we may confidently expect that chemistry, at its average, will worthily fulfil the hopes which the reasons given for its study awaken. It is in the earnest hope of contributing something, however little, to the attainment of this end by bringing together the opinions of all authorities on the teaching of chemistry in secondary schools that the following chapters have been written.

CHAPTER II

CHEMISTRY IN THE CURRICULUM

REFERENCES.

Report of the Committee of Ten of the National Educational Association. Washington, D. C., U. S. Bureau of Education. 1893. Pp. 117–123.

Woodhull, J. F. Sequence of Sciences in the Secondary School Curriculum. High School Bulletin No. 7. Albany, N. Y., The University of the State of New York. Pp. 516–523.

THE sequence of chemistry with reference to other subjects and the year in which it shall be placed are questions of great importance, since they affect profoundly the manner of the instruction and the amount that can be accomplished. This question can hardly be said to arise except in connection with the sciences. In the case of Greek the doubt lies between the second and third years. In English, Latin, or mathematics the first year is the natural place for the beginning course. In science, however, we have a choice of five or six distinct subjects which may conceivably be taught in any sequence, each in any year. Observation of schools shows that this freedom is made use of to the fullest extent. Not to occupy too much space with the discussion we may confine ourselves to the question of the order in the case of chemistry and physics, and whether the physical sciences should be placed in the earlier or later years of the high school course. Even with this restriction, there is not the slightest approach to unanimity on either of these questions, either in the opinions of school-men themselves, or in the practice of the schools. Something more final than the opinion of the individual teacher is required, since, if he is interested in securing the best work from his pupils in chemistry, he will naturally prefer to secure a place in the fourth year of the school for this work. Several of the points involved in the discussion of these questions have so important a bearing on the teaching of chemistry that we shall be justified in devoting some space to their consideration.

I. The Precedence of Physics.

Reasons for Physics before Chemistry.

a. *The Report of the Committee of Ten:* — In the report of the Committee of Ten perhaps the point which excited most discussion was the decision which they reached, that chemistry should be taught before physics. It is undoubtedly conceded by all that the logical order is just the reverse of this. The minority report of Professor Waggener states, with considerable clearness, the reasons which lead him to dissent from this part of the report. In brief, these were as follows: Since in training the pupil to make accurate observations and to draw safe inferences, the more simple subject-matter should precede the less simple, and that which is more obvious to the senses that which is less so, and since that which derives more abundant material for illustration and application from the experiences of everyday life will form a better starting point, physics seems to be indicated as the natural precursor of chemistry. He points out that a great part of physics relates to phenomena wherein the bodies concerned and their behaviour are directly perceptible to the senses at every stage of the experiment. The first results thus come from direct perception rather than by inference, and upon such phenomena the power of making inferences should first be trained. The behaviour of parts of matter concerned in chemical changes, on the other hand, is inferred, not observed; and the conceptions we form of it are less simple than those of molecular physics, since it involves a redistribution of more than one kind of matter, and the forces in obedience to which this takes place are much more complex in the matter of selection and direction than cohesion. "The rational study of chemical phenomena is therefore of a higher order of difficulty than those of physics, — certainly than those of molecular physics, a portion of the subject to which the work of the high school is largely directed." Finally he points out that chemical theory depends for rationalization so completely upon an intelligent conception of its many and close relations to physical laws that previous training in the measurement of the fundamental physical constants would seem to be indispensable.

b. *Observation in Chemistry a Study of Physical Properties:* — It appears to me that the dependence of chemistry upon physical conceptions and phenomena might fitly have been emphasized much more strongly. I think that a closer examination of the features of chemical experimentation will show this, and will incidentally point out one of the directions in which much of the teacher's effort may be fruitfully spent.

Physical Basis of Chemical Manipulation.

When any chemical operation is to be carried out, its success invariably depends upon attention to matters belonging strictly to the domain of physics. Thus, if it be a question of dissolving a salt in water, the process will take a limitless time if the solid is permitted to rest at the bottom of the vessel, along with the part of the solution which is slowly becoming more concentrated. Yet it is seldom that a pupil will spontaneously hasten the process by mixing or agitation. Again, when a gas is to be generated and collected over water, the filling and inversion of the jar of water and the displacement of the water by the gas all involve many physical questions. In more difficult experiments, particularly those connected with the determination of the molecular weight of chemical substances by one of the many simplified methods which may now be used in any high school, physical principles (vapour tension, laws of gases, adjustment of pressure before volume measurement, etc.) are almost the sole things to be considered.

A little reflection will show in a manner still more striking how, even in the study of the simplest chemical changes, the interpretation of the result depends upon a knowledge of physics. If the problem be to ascertain the effect of heating upon some body, the pupil may observe all that takes place, but, without a rapid concurrent interpretation of each feature as it presents itself by reference to physical principles, the experiment will lead to no correct conclusions whatever. The substance may melt, and the pupil must ascertain whether this is simply a physical change, or whether it involves chemical change also. The substance may boil, or appear to do so. The pupil must be in a position to distinguish between boiling and decomposition accompanied by the production of a gas, for example, by the fact that the removal of the source of heat interrupts boiling, but usually does not so promptly affect the progress of decomposition. In heating ammonium nitrate[1] we have an excellent illustration of a multiplicity of things which require physical explanation. The experiment is full of points, such as the apparent violence of the boiling while the bubbles of gas rising in the bottle succeed one another but slowly, and the cloud of smoke which sometimes accompanies the gaseous materials and passes successfully through the water, all of which require careful consideration of the physical properties of matter for their explanation.

Physical Basis of Chemical Observation.

Again, suppose that the problem before the pupil is the examination of the action of various metals upon concentrated hydrochloric acid, and that he is instructed merely in the method of bringing the materials together, and is expected to observe what follows for himself. He must have recourse to physics to ascertain whether the effect following the introduction of zinc is boiling, and consists in the evolution of steam, or is produced by the development of hydrogen. Usually his first thought is to attribute the effect to boiling, and indeed the reasoning of the observer must frequently consist in drawing a trial conclusion, and then testing it by known physical facts. Again, when the copper is introduced into the acid no action takes place. But when the mixture is warmed, bubbles of vapour are given up apparently from the neighbourhood of the copper, and the pupil is likely now to conclude that hydrogen is

Further Illustration.

[1] This illustration is fully discussed by Miss Stickney. New England Association of Chemistry Teachers, *Report of Fifth Meeting* (1899), 4.

being formed. This conclusion must be suspended, however, when he realizes that the liquid being heated is a strong solution of a gas. He must, therefore, either ascertain whether the escaping vapour contains hydrogen, or indirectly, by looking for the blue colour of a salt of copper, recognise that there has really been no formation of such a salt, and therefore there can have been no evolution of hydrogen.

It is hardly necessary to add that when parts of physics have to be drawn upon wholesale, as the kinetic theory of gases in explaining Avogadro's hypothesis and its applications, or the properties and employment of electricity in experiments in electrolysis, a previous acquaintance with dynamics and electricity is of the utmost value. In the contrary case, the extreme unfamiliarity of the whole thing interposes a tremendous drag on the progress in chemistry.

A careful consideration of any chemical experiment, even the simplest, thus reveals the fact that an intimate knowledge of the physical properties of matter is required in carrying it out successfully, and in interpreting the results. This knowledge of physics must be even more intimate than that demanded of the pupil of physics himself, for in the case of the latter the work is outlined in such a way that the subject under investigation and the method are both known beforehand. In a chemical experiment, the physical phenomena turn up without warning, and the pupil must identify them instantly and understand their whole bearing if the conclusion is to be otherwise than doubtful or hazy. In fact, the matters of immediate observation in a chemical experiment are all physical, and the data derived from these depend upon physical knowledge, and thus everything but the final conclusion is physical and not chemical.

Physical Phenomena the Language of Chemistry.

It has been remarked that "each chemical experiment is a question put to nature, and forethought and care are necessary in putting the question, and study and reflection in interpreting the answer." In view of the above we note that the chemical question has to be put in a strange language (namely, by physical methods), and the answer is returned in the same foreign language. This language must therefore be mastered before the question can be put or the reply understood. The education of a chemist consists largely in acquiring a colloquial knowledge of this language.

c. *The Conclusion:* — Whether chemistry or physics should come first is thus seen to be an idle question. Physics must come first. The question really is whether it is better to furnish a systematic knowledge of physics during the previous year, or leave it to be picked up as it is presented, hap-hazard, in the course of chemical work. When the question is put in this form, there can be little doubt in regard to the answer. It is true that the course in physics will probably not deal in any sufficient detail with some of the phenomena most intimately connected with chemistry. But the facility with which the pupil who has surveyed the whole ground in outline will acquire further knowledge of the same kind, will be incomparably greater than that of the pupil who has no "apperceptive mass" in which the fragmentary facts noted in the course of chemical work may be absorbed.

It is evident that when chemistry precedes physics, the former subject will furnish a more valuable introduction to the latter than in recent discussion has been generally admitted. A teacher of chemistry, whether he will or not, is bound to furnish some instruction in physics, and the result, while it must necessarily be unsystematic, will nevertheless assist materially in the subsequent study of the same thing. The study of either subject is bound to hasten the process of acquiring the other, but the precedence of physics is the more economical arrangement, since it will but little diminish the speed with which physics may be acquired, while greatly accelerating the progress of the pupil in chemistry.

II. Arguments in Favour of the Precedence of Chemistry.

The decision of the Committee of Ten seems to have been based finally upon the consideration that the greatest amount of mathematical training possible should be secured before the pupils enter upon the study of physics. Authorities on this subject, however, do not seem by any means to be unanimous in thinking that a course in advanced algebra and solid geometry are really indispensable prerequisites. In algebra the solution of simple equations is usually considered sufficient, while in geometry the determination of the area of the parallelogram and circle, and the volumes of the sphere and cylinder can easily be given by the teacher of physics, and thus the postponement of the work for the whole year may be avoided. It is probable that the Committee of Ten was really thinking of the value of the general discipline which these subjects would undoubtedly confer, rather than of any considerable percentage of the subjects themselves which would be required for the service of the teacher of physics.

Physics and Mathematics.

It is frequently maintained that chemistry may and should be taught more simply than physics. This is an insidious argument. Every subject should be taught simply, if by the term we mean that it should be so carefully related at every step to the previous knowledge of the pupil that over-strenuous effort on the one hand, and obscurity on the other, are avoided. But, in many cases, the simplification which makes chemistry an easy study is not of this kind. It involves not the careful bridging of all gaps and rational approach to conquest of all difficulties, but rather the mutilation of the subject and the removal of most of the science along with the difficulties. For example, in studying the action of a metal upon hydrochloric acid, we have seen that an intimate knowledge of the physical properties of the materials is required. But we may "simplify" the experiment, heading it "Standard method of making hydrogen," and direct the pupil to place zinc in hydrochloric acid. By this arrangement, as he already knows that hydrogen is a gas, no close observation, no knowledge of physics, and no reasoning are demanded of him. The whole pith of the exercise has been removed as an incident of the simplification however. Chemistry can thus be reduced to a series of cook-book receipts, and all difficulties may disap-

Is Chemistry simpler than Physics?

pear simultaneously with the removal of all the discipline which chemistry is most fitted to impart. A large part of the work may be arranged so as to consist in the formation of precipitates. Here the same set of physical phenomena is repeated time after time without variation, and the chemical conclusions, namely, that a certain substance is or is not formed, and if formed is black or red, as the case may be, may be drawn without the pupil once realizing what the physical conditions are that make this possible.

When the work has thus been conventionalized, so to speak, it ceases to deserve the name of chemistry. It has variously been designated as "cookery" and "test-tubing." Yet scorn does not seem to have much effect in appreciably reducing the amount of this kind of mechanical work. There are still some who seem to think that anything which deals with chemical materials and uses chemical terms is in some measure chemistry. There are laboratory manuals that can be used with delightful facility by the largest class, and with the least amount of supervision, which furnish the pupil with little work that is not of this description.

The mention of text-books reminds us, that the fact that high school books on chemistry appear to be simpler than those on physics, has been used as an additional support of the argument that chemistry is intended to precede physics. This really involves the question of the choice of a text-book, which we shall discuss in another chapter. It may be remarked, however, that chemistry books are not as simple as they seem. Many works intended for high school use are filled with graphic formulæ. I know no subject which is found more difficult by the beginner than the comprehension of the way in which a graphic formula represents the chemical properties of a substance. The books I refer to seem to realize this, for they make no attempt to explain the formulæ they employ. Perhaps they leave that task to the teacher. The usual result seems to be, however, that the formulæ are memorized, and are highly prized as the subjects of examination questions. If the laws and formulæ found in works of physics were to be memorized also, that subject might rival chemistry, if not excel it, in simplicity. The teacher is not compelled to confine himself to the most trivial treatment of chemistry. Avogadro's hypothesis and its consequences, involving as it does the determination of molecular weights, atomic weights, valency, and the construction of formulæ and equations, is the most fundamental principle in chemistry, and will usually be found as difficult a subject as anything in elementary physics. Professor van 't Hoff, in a recent lecture, stated that as a student he never had understood the application of this hypothesis, and that he learned it only when he became an instructor in chemistry.

Is Chemical Manipulation easier than Physical?

The argument that the manipulation in chemistry is simpler than in physics, and therefore fitted to precede the latter, is based upon the same assumption as before. If we emasculate the subject sufficiently, we can make it simpler than any other subject that may be named, but if the teaching in chemistry attempts to include the fundamental principles of the science, as the teaching in physics does, it need not suffer from lack of experiments requiring skill, patience, and knowledge. The determination of molecular weights, and the measurement of combining weights, are the most fundamental things in elementary chemistry. They are not, by any means, beyond the skill of the high school pupil, or the equipment of the high school laboratory, but they are not to be classed in simplicity with naming salts, or distinguishing 'silver,' 'lead,' and 'mercury' by the use of hydrochloric acid and ammonia.

Nor need we have recourse to experiments of a quantitative nature to furnish instances of difficulty in chemical work. The pupil in chemistry is confronted with one difficulty in every experiment, which it seems to me is not met with in the same degree in any of the other sciences. The difficulty rests on the fact that before observing, he has himself to produce that which he is to study. There is doubtless important training for the pupil who is called upon to examine a cockroach minutely, and report upon the number, location, and kinds of its appendages. But if he had to create the cockroach by a definite method of procedure, it is likely that his observations would less exactly describe the standard animal than they do. In chemical work almost every experiment will show varying results in the hands of different students, all working by the same directions. Some will use concentrated sulphuric acid or pure zinc, and so fail in obtaining hydrogen; a solution may be applied in too concentrated a form, or too much may be used; a test-tube may not have been thoroughly cleaned. Every teacher knows how puzzling the 'sports' are which the pupil may produce in this creative work. Nor is carelessness always to blame. Directions so minute as to remove all possibility of variation from the desired result would frequently be so elaborate as to be impracticable. Successful laboratory work in chemistry must depend largely on the knowledge, forethought, and skill of the pupil. The use of these is an essential part in chemical manipulation, and makes it at least as difficult as anything in the other sciences.

III. In which year of the High School Course shall Chemistry be taught?

The amount of work which can be given in a year, and the thoroughness with which it can be given, must be influenced very greatly by the general advancement of the pupil. Probably at least twice as much can be done in the fourth year as in the first, on this account alone. The necessary absence of previous training in physics, in the latter case, must greatly increase the disproportion. Every science cannot secure a place in the fourth year, and so have the advantage of reaching the pupils who are most mature and have had the largest preliminary training in other sciences. The decision must mainly depend upon whether chemistry or some other science is to be selected for most elaborate treatment. If physical geography or physiology secure the coveted position, either will obviously be benefited greatly by the fact that it is preceded by a course in chemistry. The general tendency in the secondary schools, however, is undoubtedly to emphasize most strongly the fundamental sciences, and to treat with less considerations those which are developed largely by the application of physical and chemical principles.

Argument for Chemistry in Earlier Years.

It has indeed been said that habits of neatness, care, and skill in manipulation cannot be learned after the second year of the high school, and that therefore chemistry and physics should occupy these two years. This argument, however, seems to assume that the work of the chemist requires the agility of the pianist, or the suppleness of the acrobat. Surely what is needed is rather the patient, intelligent, and forethoughtful variety of manipulation which is favoured by maturity rather than by youth. No difficulty seems to be found in training surgeons in precisely this sort of way, in years much later than those just mentioned.

Arguments for Chemistry in Later Years.

There seems to be good ground for the contention that physics and chemistry cannot give up the fullest discipline of which they are capable in the earlier years[1] of the course. Without mathematics, physics must be feeble, and without physics, the chemistry must be considerably restricted. Then, too, the continuous and minute supervision, which work in chemistry requires, must be greater the earlier it appears in the curriculum, to offset the slighter previous knowledge of the pupils. In practice, however, the much larger classes of the earlier years would entail a diminution in the supervision, rather than an increase in it, and thus still further reduce the efficiency of the work.

---

[1] Although the entrance requirements of universities should not be permitted to interfere with the arrangement of the work of the secondary schools, if their interests conflict, it may be noted that, so far as chemistry is accepted at all as an admission subject, the work done in the first or second year will almost always satisfy the very indefinite requirement. The work outlined by the University of the State of New York, and the questions asked in their examinations, are said to pre-suppose fourth year work. The requirements of the Examination Board of the colleges of the Middle States and Maryland, and those of one or two universities outside of this organization which have definitely outlined admission work in this subject, practically demand fourth year work.

Perhaps the far-reaching relations of chemistry to commerce and industry, the value of the discipline which it affords in preparing for a business career, and its importance in preparation for the study of medicine and technology are worthy of notice as inclining the schools very generally to give it the most favourable position among the sciences. It is at least certain that many bodies of recognised authority incline to recommend the placing of it late in the course. The Committee of Ten (1892) set it in the third year immediately before physics. The Committee of the National Education Association on College Entrance Requirements (1899) indicated that the last year was the most appropriate, and the University of the State of New York makes the same recommendation. In individual schools there may be good reason for departing from this arrangement. Successful curricula have been devised in which the advantage of preliminary chemistry was secured to the teachers of biology, physics, and physiography without reducing the opportunity of the pupils to secure the best training in chemistry. This is done by introducing selected parts of the subject, along with some physics and physical geography, into a course in general science which occupies the first year. The regular course in chemistry which comes later can only be benefited by this arrangement. In a few schools a compromise with the recommendation of the Committee of Ten is effected by dividing the third year between physics and chemistry, and then offering a full course in both of these subjects as alternatives in the fourth year.

Uniformity in the arrangement of the curricula of all secondary schools can never be achieved, and is probably not desirable. The chief value of the discussion in this chapter to the teacher of chemistry lies in the attempt to bring vividly before him the great importance of a clear knowledge of the physical conceptions involved in all chemical work, and the necessity which is imposed upon him, wherever his work may be placed, of making these conceptions clear to his pupils as necessity arises. Without this the work in chemistry must be mechanical and fruitless, and indeed, although dealing with the subject-matter of the science, it cannot justly be called chemistry at all.

IV. The Time to be allotted to Chemistry.

The subjects which have long been established in the curriculum in most cases run continuously through the course, and the unit of work is seldom less than a year. The sciences, however, in the struggle for recognition, have had to content themselves with a bare foothold, and in a majority of the secondary schools of this country are each disposed of in brief periods of twelve weeks. The question of the minimum length of time which may be assigned to chemistry, consistently with securing the best value for the efforts of teacher and pupil, is, therefore, one of the greatest importance.

A Full Year for Chemistry.

If the object in teaching chemistry were simply that of imparting a certain amount of information about the subject, the result would be considerable in proportion to the length of the course, no matter how short it might be. If, however, the task of contributing to the discipline of the pupil's mind is to be assigned to it, the time factor requires careful consideration. If we take into account the fact that, when a subject is taken up for the first time, familiarity has to be acquired with a new material of study, with a new language and mode of expression, and, in the case of a science, with a new mode of study by experiment in a laboratory, and a less familiar form of exercise for the reasoning powers, it is evident that much time will be consumed in overcoming the initial difficulties. In the case of chemistry eight or ten weeks at least must pass before the pupil has become accustomed to the use of a laboratory and has reached the position of being able to study the new subject in the new way. The effect of experience, which is always cumulative, is most markedly so in an elementary course in science, and even after twenty-four weeks' work, the pupil is just reaching the point at which facility in handling the subject will enable him to make really rapid progress. The influential committees which have recently reported on this subject have been unanimous in demanding at least a year for chemistry in the secondary school.

The far too common plan of teaching three sciences in a year is supported by the argument that it gives more variety, but when we consider that each of the sciences introduces a new subject, a new variety of material, a new nomenclature, new forms of manipulation, and to some extent new methods of thought, it is evident that the repeated change from one subject to another must involve a great expenditure of time on the mere machinery of each subject, and a prodigious loss of power in throwing away at each transition much that had been acquired, instead of using it as the foundation for still greater and more rapid advances in the same direction. The names of all the sciences may be included in the curriculum, but it is certain that if their number reduces too greatly the time allotted to each, the sciences themselves will never get within reach of the pupil excepting in name. If the means of the school permit the teaching of only one or two years of work in science, then one or two sciences only should be taught.

The Committee of Ten recommends that at least two hundred hours be devoted to chemistry, and that one-half of this time should be spent in the laboratory. The Committee on College Entrance Requirements of the National Education Association, the most representative educational body in this country, recommends that at least four periods a week be given to chemistry, and that half of these be periods of double length spent in the laboratory. They add that a longer time than this will be required if chemistry appears before the third year of the course. The Committee of Nine of the New York State Science Teachers' Association, in its report published by the University of the State of New York (*High School Bulletin No.* 7, 714), recommends an even longer time. If the period in the high school is forty-five minutes in length, the committee demands two double periods weekly in the laboratory, one period devoted to an experimental demonstration, two periods to prepared recitations, and suggests that three additional periods will be required for text-book and library study.

The difficulty of securing consecutive periods for laboratory work seems to be so great that particular emphasis should be placed on the importance of this. When the periods are short, experiments requiring construction of apparatus, and occupying more than a very few minutes of time in their performance, can only be accomplished under considerable difficulties. If it is found impossible to secure double periods, the apparatus may be prepared in advance by the teacher, and thus the exclusion of some experiments of fundamental importance may be avoided.

V. Continuous Courses in Chemistry.

REFERENCES.

Wilson, C. C. The Place of Science in the Preparatory Schools. SCHOOL REVIEW, VI. (1898), 211-214.

Palmer, C. S. Specialization in Preparatory Natural Science, *ibid.*, 659-671.

Arguments in Favour of Specialization.

Although the extension of the courses in chemistry in secondary schools to the length of one year has not yet been accomplished in the majority of the high schools of the country, a movement in favour of the establishment of three and four year courses in this subject has acquired such prominence that reference to it cannot be omitted. In recent articles the arguments in favour of this extension have been marshalled with such earnestness, and it must be admitted with some degree of plausibility. The disciplinary value of the old curriculum depended upon the continuous courses in Latin, Greek, and mathematics which it contained. The disciplinary value of a similar course in chemistry, or one of the other sciences, properly taught, although we have no experience of it in the secondary school period, would undoubtedly be not less than that of the older subjects. Differing in kind from these, as they differ from one another, it would be a valuable addition to the training of the pupil. It would also give a wider selection of continuous studies, and enable those who are unable to secure the greatest benefit from the classical course to get a more congenial, and, at the same time, a really worthy substitute for Greek.

Two Years the Maximum Possible.

The counter-argument that the study of science has a narrowing influence may be branded at once as preposterous. Any study, even Latin, may have a narrowing influence if taught by a narrow man in a narrow way. But this suggests one real difficulty, namely, that no non-technical or liberal course for the second or third years of chemistry has yet been worked out.[1] The real obstacle, however, in the case of chemistry, and we are not concerned with the question as it affects other sciences, is that if we agree that it should be preceded by physics, which in turn is preceded by algebra, at least two years, and more often three years of the high school course will have passed before the pupil is ready to begin the subject. Even taking the possible redistribution of the work of physics and chemistry into account, it does not seem likely that more than two years of chemistry can in any case be secured. In a few high schools this amount of instruction is given, and given successfully. The question, however, of outlining the work of the second year cannot become pressing as long as the preparation of a majority of teachers is not sufficient for a single year.

Some Favourable Opinions.

In spite of the obvious and weighty difficulties in the way of this so-called "specialization" in school science, it is surprising how rapidly sentiment in favour of it seems to be developing. Mr. Wilson mentions ascertaining the opinions of about two hundred teachers, of whom only one-third were college professors, on the question whether they preferred (a) to divide the time among four branches of science, or (b) to give the pupil a choice of four sciences or two years' work each in any two of the four sciences, or (c) to devote four years to continuous study on one subject. Only forty-three per cent

[1] This subject is discussed further in connection with that of the training of the teacher (chapter VIII).

favoured the first plan, and many of these may have done so simply because they disliked the other two still more, while forty-two per cent favoured the second, and fifteen per cent favoured the third. Of those preferring the last plan, eight or more were teachers in secondary schools.

VI. Articulation of School and College Chemistry.

REFERENCES.

Palmer, C. S. Resumé and Critique of the Tabulated College Requirements in Natural Sciences. SCHOOL REVIEW, IV. (June, 1896), 452–460.
Smith, Alexander. Articulation of School and College Work in the Sciences. SCHOOL REVIEW, VII. (1899), 411, 453, 527.

While it is generally admitted that the work of the school should be arranged exclusively with reference to the needs of the pupils of the school itself, and without reference to any special section of them which may harbour the intention of afterwards proceeding to college, there is no question but that the college has exercised a definite if subordinate influence on the evolution of the school course. In some subjects, the college has assisted in setting the pace and marking out the path which has finally been adopted as best for the pupil, whether he goes to college afterwards or not. Except in a few isolated instances, the correlation between the work of the two institutions unfortunately has been confined to languages and mathematics. In these subjects it is possible for the pupils who go to college to continue without interruption or loss of ground the studies which they pursued in school. The achievement of a similar articulation in the sciences has encountered so many difficulties that it has as yet made practically no progress.

School Chemistry a Variable Quantity.

The first difficulty lies in the extraordinary diversity in length and in content of the courses in the same science in different schools. In chemistry, the time varies from twelve to forty weeks, and the instruction may be entirely in general chemistry, almost entirely in qualitative analysis, or it may dispense with the laboratory. The college, drawing its freshmen from a hundred different schools, cannot furnish a course which will fit equally so many differing foundations, and it does not attempt the task. President Eliot says, "It would be a pity if we could not adapt our courses in college to any good teaching in the schools." If Latin and mathematics, however, had remained one-tenth part as full of divergencies as school chemistry, the present system of co-operation would never have been brought about. It is difficult to believe that chemistry possesses any property which makes this divergence unavoidable.

Attitude of the Colleges.

The second difficulty in the way of articulation is the considerable diversity in the elementary courses of different colleges, and therefore in the work, part or all of which, pupils in the same school in going to these different colleges must attempt to anticipate. A third difficulty is that many colleges give no admission credit for chemistry,[1] and the

[1] The preliminary report of the Committee on College Entrance Requirements of the National Education Association, published in the SCHOOL REVIEW, IV. (June, 1896), 341–412, gives some startling information in regard to this subject. Of the fifty-six colleges and universities whose admission requirements were discussed, only thirty accept chemistry at all. A further study of the relation between admission and college chemistry in these thirty institutions, which I had occasion to make and have fully discussed in the SCHOOL REVIEW, VII. (1889), 411, 53, 527, shows that only three have definite entrance requirements, and provide a definite mode of handling those who offer them. A dozen or so place the students who offer chemistry into the college course along with beginners, and the remainder seem to attempt a rough sifting by which the better prepared students go into advanced work, and the less well prepared into the elementary course.

Professor Bardwell of the Massachusetts Institute of Technology presented to the Sixth Meeting of the New England Association of chemistry teachers some facts which illustrate the method in the last class of institutions. In the autumn of 1899, one hundred and fifty-eight students offered chemistry for admission to the Institute, being 50.3 per cent of the total number entering. After five weeks eighty-six of these students remained in advanced courses, while seventy-two retired voluntarily into the elementary course. It is evident that a majority of the eighty-six were most likely only partially fitted for the work in which they found themselves, while the seventy-two were all misfits in the elementary course, since they had all studied more or less of it before. It is evident from

rest, with few exceptions, give credit for anything that is presented, and thus make the arrangement of a logical sequel to high school work in the subject within their own walls impossible.

The few universities which insist upon a definite amount and kind of chemistry do not agree at all in regard to the kind, and thus when the school seeks the advice of the college, as it often does, the utterances of the latter in regard to chemistry lead to nothing but discouragement and distraction. The Committee of Ten reported definitely "that there should be no difference in the treatment of physics, chemistry, and astronomy for those going to college and scientific school and those going to neither." The principle would have been something more than a mere *doctrinaire* statement if it had read, "When the secondary schools have decided upon the length, aim and content of their course in chemistry, all colleges should accept this for admission." [1]

The fourth difficulty in the way of articulation is that no advancement is granted to pupils who offer chemistry as an admission subject. I have elsewhere discussed this subject more fully, and may be permitted to quote a few lines.[2]

"The college should grant advancement in the series of its courses in each science to an extent corresponding to the admission credit given. In other words, it must recognise adequately, and in a practical form, the extent to which the school work may fairly claim to constitute an anticipation of its own.

"To effect this, each department in the college must adapt its own courses so that one of them shall offer a suitable continuation of the preparatory work. This will be open to those

---

this that the Institute has no definite requirement in admission chemistry, and must, like other institutions of the same class, share with the schools the blame for this chaotic state of affairs.

[1] Arguments similar to the above, and leading to the same conclusions, have been urged most strongly by the Committee of Nine of the New York State Science Teachers' Association in its first report (University of the State of New York, *High School Bulletin No.* 2, 478–480).

[2] From the SCHOOL REVIEW, VII. 456.

students who enter with a credit in the subject, and such students should never be required to begin the science over again in the same class with those who lack this credit and preparation."

The course in continuation of school work will not usually be the second regular college course, for the school work will not be the equivalent of the first course in college. When the college, as it often does, attempts nothing beyond a secondary school course in elementary chemistry, it deliberately throws away the advantages which the more rigid selection of its students, the smaller size of the classes, the greater maturity of the constituents of these classes, and the greater amount of work which can consequently be demanded of them, place in its hands. The college introductory course should be heavier by at least a half, and a distinct class should be formed for those who are not beginners and desire a sequel to secondary school chemistry. This arrangement should certainly be possible in the larger universities, and especially in technical and medical institutions in which all the students are required to study chemistry, and in which, therefore, a sufficiently large number will have offered it for admission to warrant the formation of a separate class. Where no proper sequel is offered, and chemistry inside the college is optional, the pupil takes an elementary course, in which much that he has already studied is repeated, of his own free will. Where the pupil is required to take college chemistry, however, and admission credit has been granted, the institution is under an obligation to furnish fit instruction to the candidates.

The College must offer Two Independent Courses.

The movement in favour of unity in the matter of secondary school chemistry will doubtless be materially assisted by the recent inauguration of a college entrance examination board, by the Association of Colleges and Schools of the Middle States and Maryland, and the preparation by it of syllabuses[1] of admission

[1] The requirements in all subjects may be obtained by transmitting the price, ten cents, to the Secretary of the College Entrance Examination Board, Sub-station 84, New York, N. Y.

work in all secondary school subjects. The syllabus in chemistry is based upon the report of the Committee of the National Educational Association, and will probably be accepted not only by the universities within the association, but also by the great majority of the institutions of learning in the country.

CHAPTER III

THE INTRODUCTION OF THE SUBJECT

BIBLIOGRAPHY.

Richards, T. W. Requirements in Chemistry for Entrance to Harvard College. Cambridge, published by the University (1900). Pp. 4–10.

Freer, P. C. The Teaching of Beginning Chemistry. Proceedings of the National Educational Association, 1896. Reprinted in SCIENCE [N. S.], IV. 130–135.

Smith, Alexander. The Value of Chemistry. Proceedings of the National Educational Association, 1897. Pp. 945–951.

Report of the Committee of Nine of the New York State Science Teachers' Association. High School Bulletin No. 7. Albany, N. Y., The University of the State of New York, 1900. Pp. 708–721.

I. Impediments to be overcome or avoided.

Study of Language and Science Contrasted.

WHILE the introduction of any new subject must of necessity be difficult, there are special reasons which make the demand for unusual tact and skill on the part of the teacher of science imperative. The introduction of a new language, for example, does not present the same degree and kind of difficulty. The pupil has been accustomed from his infancy to handling the problem of words, their meaning, and their relations, and there is no novelty in the material, or, to any great extent, in the method. The operation of noting the usage of words, for example, and determining their precise significance, " the formation of hypotheses . . . and repeated modification of hypotheses after they have been brought to the touchstone of experience," and, in general, the operation of organizing isolated facts into knowledge, which Professor J. G. Macgregor has styled knowledge-making, has, in the direction of language, become a habit. Much of this work may have been unconscious, but it has none the less resulted in education with especial application to a particular kind of problem. The objects of the material world have not been studied with anything like the same care, for attention to physical matters has not occupied almost every waking instant, and there has not been the same inexorable necessity for minute and exhaustive organization of the phenomena which they present.

Then, too, the study of language furnishes an endless succession of simple problems in which the same forms recur at short intervals an endless number of times. A science, on the other hand, presents "problems with a greater range of difficulty on a material which is in general more complex."

Science more Difficult.

Furthermore, when a new language is presented, the assistance which the pupil receives from the grammar and dictionary has no parallel in scientific work. The contents of these aids to study are classified in such a way that the problem of ascertaining the meaning of a word or phrase can be at once reduced within certain narrow, clearly defined limits. The laboratory directions, indeed, attempt to instruct the pupil how he shall himself produce that which in science takes the place of the text, the phenomenon to be studied. But unless these directions play the part of an interlinear translation also, he has to provide from his own previous experience the ability to separate the significant from the insignificant factors among the many details he may observe, and to furnish, upon the same presumably rather meagre basis, the correct interpretation. The teacher always has it in his power to simplify the problem by affording guidance, but, if this is carried too far, the benefit of learning from experience under conditions which far more closely resemble those of actual life than is the case in language study, is snatched from the pupil's grasp. Acquiring the ability to make knowledge is education, and to shield the pupil from the necessity of doing this with the material which science supplies, is to deprive him of that element in his training which science is in an especial degree fitted to furnish.

Not only, however, does the beginning work in a science present an unfamiliar material for study, but it should seek to cultivate an attitude which is for the most part entirely new. The work in chemistry can be made almost wholly inductive in method, and must be made altogether so in spirit. The pupil encounters an additional difficulty in the acquired mental habit which he has of developing consequences by speculation. It is the hardest thing in the world to compel him to stick closely to the facts and to test such inferences as he may draw by renewed scrutiny of the data, and perhaps the performance of new experiments, before adopting them. The symmetry of an idea, and its logical harmony with conceptions already existing in his mind, blind him to the fact that a dozen competing ideas might have arisen in the same connection, and yet none of them be confirmed by experience. In geometry he is accustomed to the developing of a system from a few simple conceptions, and he has still to learn that in science a multitude of facts are required for the foundation of one conception. Not only does the pupil suffer from this difficulty, but the teacher himself may follow the lines of least resistance, and, allured by the rapid progress the pupils make, conform his teaching to methods to which they are accustomed, and so throw away the opportunity of making a new start which the study of a science furnishes. He may thus all too easily pervert it into a continuation of the same kind of discipline, instead of making it the starting point of a new one. The teacher must be continually on the watch lest defects in his own training, which he has not later observed and remedied, lead him to teach chemistry as a dogmatic system of principles with which the concrete experience in the laboratory has little more than a nodding acquaintance.

Inductive Spirit in Science.

Still another feature of chemical work which in some ways forms an impediment to the beginner, is the attitude of an original observer in which he is to be placed. This attitude is a strange one to him, for he has been accustomed to accept facts from books or his teacher as the basis of his work, and even to derive most of his opinions from sources other than his own intelligence. It is difficult if not impossible to conduct the elementary instruction in a language in a way which will have any other effect than to confirm the mind of the pupil in this attitude. Before beginning a science, therefore, he has acquired the habit of relying upon authority for most of what he learns. It is the special boast of work in a science, that, as it proceeds, the pupil is bound to see that the facts may be derived from his own observation, and the conclusions may be drawn by his own unaided efforts. It is held that scientific work thus furnishes an exercise in independent thought much more readily than the study of language.

Self-reliance in the Laboratory.

The number and subtlety of these pitfalls to which the introductory work in chemistry is especially liable, make it important that we should devote a chapter to the discussion of the most natural method of approaching the subject, and of the principles which should first be the objective of the instruction.

II. What Phenomena shall furnish the Basis of the Introductory Work.

The course in chemistry frequently begins with a part which is intended to be introductory, and is not a portion of the systematic presentation of the subject. There must of necessity be some attempt during the earlier part of the course to marshal before the pupil the various types of chemical change, the most characteristic features of chemical action, and the constantly necessary habits which he should form in doing chemical work. In this chapter we shall not attempt to elaborate any novel method of approach. We shall simply seek to decide which are the most important generalizations, and how they may be brought to the knowledge of the pupil. In concrete form our conclusions will be found embodied in many of the available text-books. The common statement of the nature of the subject-matter of the science, which is usually to the effect that chemistry deals with the changes in composition which matter undergoes, and with the accompanying physical phenomena, will furnish us with a starting point.

a. *Classification of Various Principles of Arrangement:*—The elementary study should clearly begin with familiar forms of matter, and familiar phenomena should be selected. If any of the earlier facts are unfamiliar, they must at least be closely related to those which are familiar. Then, also, the selection must consider the facility with which the phenomena can be subjected to experimental study by one who is as yet untrained in the methods of the science. Thus while the action of soap upon water, and the effects of the solution produced by it upon the dirt, are exceedingly familiar, they are not capable of simple experimental investigation. Finally, the chemical changes studied must be of a simple nature in the chemical point of view, since then alone will they form an easy vehicle for the passage from the realm of simple fact to that of chemical knowledge.

The Earliest Observations.

At this point a divergence takes place which enables us to classify the ways of treating the subject roughly into three kinds, and, it may be remarked, imposes upon us ultimately the necessity of deciding which is more applicable to the case of any given set of pupils. It will be noted that we are speaking at present mainly of the ways of selecting the content, and not of modes of presentation, inductive, deductive, or otherwise. One method proceeds by selecting from common materials those whose general physical properties must be familiar even to the youngest, namely, solids, and frequently devotes a very considerable amount of attention to quite a series of studies from which work with gases is, as far as possible, excluded. Another variety of treatment deals indeed with familiar substances to begin with, but does not restrict itself to the most familiar materials physically. In fact, it deliberately leads up as rapidly as possible to the properties of air and the chemical effects of oxygen, pursuing its way after that largely through the study of other gases.

The Three Principles of Arrangement.

When the former of these methods is employed, the main object is to put the pupil in the attitude of a discoverer. The problems are selected therefore, not because of their chemical importance or their relation to the development of an organized knowledge of the science, but solely because they are simple, since thus alone is there any hope of realizing the object in view with any degree of completeness. The facts as they are accumulated lend themselves easily in this less systematic study of the subject to the construction of the ordinary generalizations of the science. But the ultimate results come more slowly than they would with the more systematic treatment.

Nature Study Method.

Largely different must be the arrangement of the work, if the most logical presentation of the framework of the science is to be made one of the objects. While the same methods are pursued in matters of detail, this plan seeks, as directly as possible, to reach the means of explaining the basis of our modern mode of expressing the quantitative relations involved in chemical change. In other words, this plan handles the gases as soon as possible in order that it may quickly lead up to the explanation of Avogadro's hypothesis and the consequences which follow from it. Until this hypothesis has been discussed, everything else relating to the appropriate statement of quantitative relations must remain largely in suspense, unless we are willing to teach these matters in an empirical manner without examining their basis or knowing the centre from which they are controlled. When the attempt is made boldly thus to grapple with the foundations of chemistry, the pupil must perforce be brought rapidly through the most indispensable stages leading to the study of chemical change in the light of Avogadro's hypothesis. His early work must thus deal largely with gaseous materials, and the pedagogical advantage of greater familiarity which solid bodies afford must be sacrificed.

Theoretical Method.

Still a third method, which, however, is closely related to the last, may be defined. In this arrangement of the material the desire is, as rapidly as possible, to bend the order of study into a series of chapters dealing with successive elements, arranged in an order something like that which, in spite of slight variations, is in its general features common to most books. The second method was an arrangement with reference to theory; the third is an arrangement with reference to chemical materials, with the theory distributed at convenient intervals. The order here seems to be determined in the first place by a desire to conform to the historical development of the subject. Oxygen, air, and water thus find an early place. This *motif* presently gives place to the impulse to arrange the elements in accordance with the natural families.

The Historico-Systematic Method.

Each of these methods of arrangement, the nature study, the theoretical, and the historico-systematic, has its merits. The decision as to which is more suitable will depend largely upon the advancement of the pupil whose instruction is under consideration. The first method is practically that which is adopted in nature study, excepting that it may be expanded beyond the limits of the familiar materials to which the latter is confined. It is applicable to the youngest scholars, and in general would probably be the best arrangement for pupils in the first year of the secondary school. The two latter methods, suitably modified by importations from the former, might enter largely into a course given in the later years of the secondary school, especially if the pupils had already studied physics. Their greater maturity, as the result of continuous work in languages, mathematics, and physics, would more than offset the more rapid progress they would be called upon to make through unfamiliar ground. The more highly developed intelligence required for the successful accomplishment of the more difficult task should be by this time available.

b. *Various Arrangements Illustrated by Reference to Existing Text-books:* — It is so important that the teacher should have a clear idea of what is implied in the arrangement of the text-book which he may adopt, and of the precise demands which this will make upon his pupils, that it may be well to indicate books which will furnish examples of each of these methods of handling the subject.

The Nature Study Method.

The best example which I know of a work of the first kind is *An Introduction to the Study of Chemistry* by Professor Perkin of Owens College and Dr. Lean[1] of The Friends' School, Ackworth. The study of chemical change begins (p. 126) with an experimental examination of common salt, chalk, sand, washing soda, iron pyrites, and other common materials. This is followed by a study of the common acids, and common alkalies, and this again by the relation between acids, bases, and salts. The next topic is the 'fixed air' of Black, followed by rusting and combustion. The remaining subjects, which are not numerous, need not be given. There is no attempt to develop the science in a conventional manner. The way in which ordinary knowledge of familiar materials is gradually transformed into scientific knowledge of the same things is worthy of careful examination.

Aside from this, which is our main reason for mentioning the book, it presents other features which make it exceedingly instructive. The first part (up to p. 125) deals entirely with physical properties. This is doubtless done in recognition of the fact that the pupils using it will be entirely ignorant of physics. The selection of material, however, is naturally not that which the physicist would make in presenting his subject symmetrically, but shows rather the parts of physics particularly important in chemical observation. We shall revert to this subject presently. The reader accustomed to the decoration of the pages of every chemical work with numerous equations, and supposing that these are indispensable parts of the science, will be surprised to find that in a course of this kind they may be dispensed with entirely without appreciable loss in clearness. Then, too, since the work does not pretend to be a treatise on chemistry, it gains the unquestioned right to omit what it

---

[1] Perkin and Lean. *An Introduction to the Study of Chemistry.* London and New York, Macmillan. 1896.

pleases, and thus shows that much chemistry of a perfectly sound description may be taught without a single mention of atoms, molecules, or valency. While the presence of these is doubtless demanded in a book treating the subject by the second or third method, a study of the aspect which chemistry presents in their absence will prove exceedingly instructive to any who may think that chemistry begins with these conceptions, and it is to be feared sometimes act as if it ended there also.

Method used in Great Britain.

Much of the recent discussion of the teaching of chemistry in Great Britain has been concerned with urging the moulding of instruction in the subject on the lines of that method of the three which we are now discussing. A syllabus of elementary chemistry published by the Board of Education,[1] described as the 'alternative elementary stage,' furnishes another instructive example of what we have called the nature study method.

A Committee of the British Association[2] suggested a plan of study closely resembling those we have just mentioned. The new Syllabuses[3] issued by the Incorporated Association of Head Masters also present a well-devised and thoroughly tested course of a similar kind.

The Ideal in American Secondary Schools.

As we have suggested, the plans of the first kind are not accepted as the basis of such work as is usually attempted in the later years of the secondary school in America. They do not present that connected and complete account of the subject which in these years is generally demanded. Pupils trained with their assistance would have a sound knowledge of chemistry, so far as that subject had been covered, but they would not be able to pass

---

[1] *Directory, with Regulations for Science and Art Classes.* London, issued annually by the Board of Education and sold by Eyre & Spottiswoode. The *Syllabus of Chemistry* may be had separately.

[2] For references and further discussion of the nature study plan, see Heuristic Method, Chapter IV., Section IV. (p. 105).

[3] *The Elementary* (1900) and *Advanced* (1899) *Syllabuses* are published by Whittaker & Co. (London).

examinations for admission to most colleges, since they would probably know nothing of equations or the atomic theory. Their work, in spite of its excellence, would lack some of the conventional signs which usually mark a knowledge of chemistry, and sometimes take the place of it. The study of them, however, will afford to the teacher a valuable demonstration of the application of pedagogical principles to the study of chemistry. The teacher fresh from the college or university, especially if he has been highly trained in chemistry, is apt to have forgotten the almost innumerable steps by which he reached his knowledge of the science, and to give his pupils work which assumes this knowledge rather than instruction which will confer it. Under such circumstances their acquisition of the subject becomes purely mechanical, and, in the highest sense of the term, wholly uneducative.

Theoretical Method.

The second of the three guiding principles in the arrangement of the introductory work, the theoretical, is illustrated more or less clearly in a number of books. The idea is typically presented in Dr. Torrey's *Elementary Studies in Chemistry*.[1] When he reaches the first chemical experiments (p. 62), after introductory work dealing exclusively with physical properties, he proceeds rapidly, through the study of combining proportions by volume, with water and hydrogen chloride as the concrete materials, to the statement of Avogadro's hypothesis (p. 101) and the consequences which follow from it. The intervening matter, while it does not take the shortest course possible towards this goal, nevertheless is lightened of much of the material usually treated in connection with the chemistry of oxygen, hydrogen, and water, in order that the development of the theory may not be impeded. After Avogadro's hypothesis has been disposed of, a larger proportion of general chemical work begins to appear, while at the same time formulæ and atomic weights are discussed. Almost all of the study of the properties of the elements and their compounds

[1] Joseph Torrey. *Elementary Studies in Chemistry.* New York Henry Holt & Co. 1899.

thus follows the theoretical matter. The recent work of Professor Young[1] is arranged similarly on the same general principle so far as the theoretical part is concerned. The facts employed up to the end of the development of the theory (p. 89) are selected at random from various parts of the subject, and in consequence solely of the readiness with which they furnish experimental support for the theory. The systematic treatment of the elements then follows. Professor Freer's elementary work[2] resembles these books in placing the logical development of the theory in the foreground, with the employment of a minimum of selected facts, and differs from them only in that the treatment of the rest of the science is much briefer, and the whole ground covered much less extensive.

The Historico-Systematic Method.

The third principle which has been mentioned as affecting the arrangement of the work, the historico-systematic, is commonly associated with the second, and the presentation of the elements one by one is usually found in combination with, and as a modifying factor in, the application of the second principle. The well-known books by Professor Remsen[3] and Dr. Newell,[4] however, illustrate very well the continuous development of the principles along with an arrangement of the material in the normal order. The two run side by side, and the more theoretical portions are taken up at convenient intervals without any effort to introduce each at the earliest moment at which this is theoretically possible. Professor Newth, in his *Elementary Inorganic Chemistry*,[5] pursues essentially the same plan. The connection between experiment and inference is worked out systematically with admirable

---

[1] A. V. E. Young. *The Elementary Principles of Chemistry*. New York, D. Appleton & Co. 1901.

[2] P. C. Freer. *The Elements of Chemistry*. Boston, Allyn & Bacon. 1895.

[3] Ira Remsen. *Introduction to Chemistry* (*Briefer Course*). New York, Henry Holt & Co. London, Macmillan. 1893.

[4] Lyman C. Newell. *Experimental Chemistry*. Boston, D. C. Heath & Co. 1900.

[5] G. S. Newth. *Elementary Inorganic Chemistry*. London and New York, Longmans, Green & Co. 1899.

clearness. The theory is introduced at suitable intervals without haste and at the same time without undue delay. The report of the Committee of Nine of the New York State Science Teachers' Association[1] contains a detailed outline of introductory work, in which the same combination of these principles of arrangement is observed.

The combination of the second principle with the first may be seen in the late Professor Cooke's *Laboratory Practice.*[2] Ordinarily the theory is consistently developed after certain physical properties and manipulations have been studied, but here the rate of its development is modified by the effort to combine with this much experimental work on a variety of materials. The difference from the treatment last described lies in the fact that this chemical experience is afforded, not by the more or less systematic study of the elements, but by handling a number of miscellaneously selected topics.[3]

c. *The Present Ideal of the Secondary School Course in Chemistry :* — Of the various plans outlined above, each is admirable in its way. Each has advantages for certain purposes. It remains for the teacher to decide which is most likely to suit the case of his particular set of pupils. Nor need the choice of a book wholly determine the kind of instruction to be given, although it must influence it largely. At present the prevailing tendency in American secondary schools seems to be towards the use of the historico-systematic, if not the theoretical style of book. At least the recent works seem to

Prevailing Ideal.

[1] *High School Bulletin No.* 7. Albany, N. Y., The University of the State of New York. By post 35 cents. This bulletin contains so much suggestive matter of the highest interest to the teacher of chemistry, aside from this report, that the reader should not fail to obtain it.

[2] J. P. Cooke. *Laboratory Practice.* New York, D. Appleton & Co. 1891.

[3] Amongst the other elementary text-books whose methods are worthy of study are: J. E. Reynolds. *Experimental Chemistry for Junior Students;* Part I., Introductory; Part II., Non-metals; Part III., Metals. Longmans, Green & Co. 1897. J. Walker. *Elementary Inorganic Chemistry.* Bell & Sons. 1902. Henry Roscoe. *Lessons in Elementary Chemistry.* Macmillan. 1890. Storer and Lindsay. *Manual of Chemistry.* American Book Company. 1894.

emphasize these conceptions in their selection of material and method of arrangement. Observation of the work of many schools confirms this belief. It is evidently a prominent aim of the teacher at the present day to give his pupils a well-rounded account of chemistry as a science; to give him a bird's-eye view of the science as an organized system of knowledge; in fact, to show him an outline plan of the results of the science as they appear to the chemist himself. With this purpose in view, Avogadro's hypothesis and its consequences, formulæ, and equations (often even graphic formulæ), and many somewhat artificial experiments with strange substances have to play a conspicuous part in the instruction.

Treatment, Academic and Formal.

There can be no question of the value of a well-ordered outline knowledge of the whole science for the understanding of its parts. But an outline or map of an extensive territory is not an end in itself. The sketch is chiefly useful to those who know the country first hand and can fill in from their own experience the detail of many parts and so form a true appreciation of the whole. It is the knowledge of this detail which constitutes a genuine acquaintance with the subject of the outline. It is thus unfortunate if the effort to give a systematic plan of the subject is allowed to occupy the foreground, while the method and detail of the science are suppressed by lack of time. It is important therefore pointedly to call attention to a possible danger in this direction. The extensiveness of the field covered by the book tempts one to make a superficial rush through the whole subject instead of taking time for a detailed study of the fundamental things of the science. Thus the hypothesis just mentioned is indeed fundamental, in a sense. But behind it and on every side of it there is something without which it is merely an empty phrase, and that is the ability to understand and reason about chemical problems as they present themselves in the laboratory. The behaviour of concrete substances and mixtures of substances in real test-tubes and flasks and the phenomena around us in

Course often covers too much Ground.

nature must be understood before more abstract matters acquire any significance.

The point is that there are few, very few, secondary schools in which the time allotted to chemistry permits treating both of these aspects adequately. I hesitate to quote experience in higher institutions in discussing school work. What a university student can do, in many cases the school pupil cannot do. Yet, what a university student cannot do, with the advantages of maturity and preparation which he possesses, must surely be, *a fortiori*, beyond the pupil in the high school. I find that more than a hundred hours (of 60 minutes) in classroom and laboratory are required for the introduction to the subject. In the course of this, a beginning, and only a beginning, is made in learning to observe; the theory up to and including Avogadro's hypothesis is taught; and a few elements and compounds are studied in a very elementary way. All that can be covered in this time, besides introductory matter, is the chemistry of oxygen, hydrogen, water, chlorine, hydrogen chloride, air, and nitrogen (the element only). It would be desirable to give even more time to this seemingly meagre programme. Indeed more would be given if the imperative necessity of covering the whole subject in outline within a total of 325 hours did not compel the adoption of a more rapid gait at the expense of thoroughness. Now 100 hours is half or more than half the whole time allotted to the subject in most secondary schools. In giving the same extent of work with the pupils of a secondary school I should be compelled to occupy an even longer time, and all hope of covering the subject would have to be given up.

The Intensive Method.

More rapid progress through the science can be had only by substituting the memorizing of results for genuine study of chemical problems. Learning chemistry as it is and making a rapid survey are two things which, under the conditions imposed by the programmes of secondary schools, are incompatible. What is meant by genuine study of chemical problems? Let us take an example. Suppose we learn that hydrogen burns in air and forms water, and illustrate the fact by burning a jet of hydrogen in the laboratory and holding a cold beaker over the flame. This exercise has the appearance of having taught the fact, and we go off in the belief that the pupil thoroughly understands all about it. If, however, we overhaul his conceptions, we soon find that we have conventionalized the experiment and the result has been learned as a purely mechanical acquisition.

Illustrations.

To test this statement, take the class after this exercise, extinguish the jet of burning hydrogen, hold a cold beaker against the jet of unlighted gas, and ask what the class thinks of the moisture which the gas is seen to deposit even in this condition. In a large class a few will suggest that the water comes from the union of the hydrogen with the oxygen of the air, — showing that they have failed to appreciate the significance of the *lighting* of the jet. Others will think it comes from condensation of moisture in the atmosphere, although they can give no explanation of how this happens. I have not yet encountered a class of beginners in the university in which a single member is to be found who can suggest the correct explanation, so little are students, even well-prepared and intelligent ones, able to apply the knowledge of physics they possess. Further questioning shows that, although the whole preparation of the experiment had previously been done by the pupils themselves, and attention had been drawn to the heat developed by the action, not one realized spontaneously that his flask contained a liquid which was warm and consisted, to the extent of 80 per cent. of water through which the hydrogen was passing. The bare skeleton of the action, as it appears in the equation $Zn + H_2SO_4 \rightarrow ZnSO_4 + H_2$, seemed to be all that their minds had consciously grasped of the whole paraphernalia of the action.

It is only after a thorough discussion in which attention has been called to the details, that the pupil realizes that the first condensation of moisture was quite inconclusive as a proof that water was formed by the union of hydrogen and oxygen, sees the need of drying the gas, and finally learns that chemical work includes far more and far more important things than putting together the materials stated in the equation. Now this sort of experience can be duplicated in almost every action studied, and it must be constantly repeated in a thousand forms before any intelligence about chemical work can be developed. But this sort of work takes time which might otherwise be devoted to passing on to the acquisition of a quasi-knowledge of new actions.

It is doubtless assumed by the writers of the systematic variety of treatises, that the thorough development of all sides of each experiment will be brought out by the teacher. They furnish the skeleton, and the teacher does the heavier share of the work. The most of the real chemistry is between the lines. So, as has been said, the character of the book need not determine the character of the instruction. A skeleton book need not lead to an attenuated, fleshless, academic treatment of the subject in the class. But the inexperienced teacher, and it is for him chiefly that this is written, will find that it takes much thought and experience to extract the meat from the work, that the skeleton in the book may be fitly clothed withal, and that there is a constant temptation to treat the skeleton itself as if it were the chief dish of the feast.

Why is the Chemistry in Schools so Formal?

There must be some reason for the tendency to make school text-books of chemistry more and more academic. It must be that some essential element in secondary education would be lacking if the formal survey of the science were not conspicuous, or its prominence as a characteristic of school chemistries would not be so great. Perhaps the key may be found in the use to which the chemistry, or the training it furnishes, is to be put in after life. This test ought to furnish the explanation of all peculiarities of secondary school work. Some of the pupils go to college, some become teachers in grammar and grade schools, the great majority go into the affairs of business or professional life. What purpose can formal chemistry serve for each of these classes?

To the last a knowledge of formal chemistry, which is not

also much more than this, cannot be of great use. On the other hand, a quick perception of conditions, ability to reason surely about the causes of physical phenomena, capacity to study materials and to devise means of accomplishing definite ends with them and, in the broader view, a confirmed habit of getting to the bottom of everything that is observed, will be invaluable. It will be of far more use than an array of picked information which all soars a little above the plane of experience as we find it, and has not been brought down to the every-day level.

Chemistry in Preparation for Life.

Again, the teacher in the lower schools gives instruction in elementary science. But here a formal knowledge of chemistry only hampers her, unless she has happened to have the opportunity to go further into the subject. The chemistry of the grades, so far as their nature-study work can be said to include chemistry, must not be consciously chemistry at all. A knowledge of three distinct ways of making chlorine and the ability to name the fundamental laws of chemistry will not be of much assistance. The study is approached from an entirely different side. For example, what are leaves made of and where do they go when they disappear every autumn?[1] The high-school text-books do not tell us how to lead the child to some comprehension of this latter wonderful natural fact. We may take two equal heaps of leaves and weigh one and keep the other for reference. Then we dry the weighed leaves. For small children we condense the moisture that comes off and show it. Thus we have a certain weight of dried vegetable matter and a certain weight of water, both of which can be handled and compared with the undried specimen. Then we burn the dry leaves and get a certain weight of ash, and a loss in weight represented by the burnt material. Then we treat the ash with water, and part dissolves, and is recovered by evaporation. Part remains insoluble. Thus we lay bare the nature of leaf material. The child sees

Chemistry in Preparation for Teaching.

[1] From Jackman's *Nature Study for Grammar Grades* (Macmillan), chapter VIII.

also that the ash part came from the soil, the water from rain or other moisture of the soil. The subject can be pursued further if it seems desirable. My point is that while intelligent teaching of this sort of thing will be assisted by a systematic knowledge of chemistry, it will be quite impossible if the knowledge of chemistry was of the formal, equation-loving sort and nothing more. Evidently the too academic variety of chemistry is not intended to help the teacher in the grades.

Chemistry in Preparation for College.

Finally, can it be that the teaching of chemistry in secondary schools has been modelled to suit the supposed need of the pupil who is going to college, and that the needs of the vast majority who do not go to college are in danger of being sacrificed? It would be a pity if that were the case. There is no reason why the college should demand a knowledge of formal chemistry for admission. Any good chemical instruction which really teaches chemistry of some kind should be gladly accepted. Of course a knowledge of how to reason and how to work intelligently is difficult to test by examination, while the presence of a knowledge of formal chemistry can be readily ascertained.[1] But if the intelligent knowledge of chemistry is more valuable for other purposes, and at least as valuable to the college entrant, it is the business of the college to find some way of testing it. It is much to be feared that the college ideals of what constitutes a

[1] The various existing reports describing courses in chemistry for secondary schools are perhaps in part responsible for perpetuating the impression that, whatever else is desirable, a formal survey of the science is indispensable. It is relatively so easy to give a brief yet comprehensive list of topics, and so difficult to describe effectively the spirit and manner in which the instruction is to be carried out, that the former never fails to put in an appearance, while the latter is slighted. The list may be suggestive and valuable, but if it is unaccompanied by a full discussion of how the teaching is to be done, occupying a space which, to represent the relative importance of the two parts, would have to be ten, or even a hundred times more extensive, the impression conveyed by the report as a whole must be misleading. The fact that preparing and securing the passage of such an extensive report are well-nigh impracticable is the only excuse that can be found for the way in which the crucial part of the task has hitherto been avoided.

genuine knowledge of chemistry, which are themselves much in need of reformation, have been permitted to influence the kind of chemistry taught in the schools to far too great an extent.

In what has been said above there is danger that I may be misunderstood. No one can doubt the pre-eminent value, the absolute indispensability to the chemist, of a systematic view of the science. But an attempt really to give this view in any genuine way, when the basis is lacking, must be futile. Along with this view, the chemist has also at his command all the details which have gone to the making chemistry in the past and which make chemistry for him a living reality. I am therefore raising the question whether putting the broad sketch of the whole in the foreground, and leaving the details to the teacher and to chance, is not a reversal of the proper order. Some of both must be given. But the technique of experiment, observation, and induction, and the habit of using it come first.

It was with thoughts of this sort in mind that emphasis was laid on books of the nature-study variety. Much of their spirit may be infused[1] into teaching which professes to follow one of the other plans. Adaptation to the point of view of the beginner will demand such an infusion, if the instruction is not to be altogether artificial. We cannot approach a class with the idea that here is a certain outline of work to be done, which has been selected because its importance is evident to the mature mind of the chemist, and regard the pupil as being there to receive the dose.[2] The case is rather that the pupils are there to be educated and assisted in development, and the work must be adapted to their preparation and needs.

At the same time this book has not been written to advocate a new kind of chemistry, but to assist the teacher who is giving the kind of chemistry which is at present demanded. So we shall assume for the most part in what follows that one of the

[1] See the paragraph on the attitude to be cultivated in the pupil under Instruction in the Laboratory, chapter IV., section IV. (p. 105).

[2] See May M. Butler, SCHOOL REVIEW, X., (1902), 52.

more or less systematic texts is being used, and simply try as occasion offers to show how the instruction based upon it may be adapted to give the most benefit.

To sum up our conclusions, the study of chemical change, the generalization of the features which it presents, and the acquisition of the habit of applying the knowledge thus acquired, are the business of the student of chemistry. In assisting him to a mastery of the science, the first thing is to lay a solid foundation in the knowledge of the detail of observation. The second thing is to lead him up to generalization, for without this the work will not be scientific. The third thing is to exercise him in application, otherwise the work will not be useful. Ordinarily the work is controlled in the fourth place by an effort to approach the systematic arrangement of the elements, or at all events, sooner or later, to reach this. Finally, it is undoubtedly useful to combine these objects with the presentation of much of the early matter in the historical order.

Conclusions in Regard to Introductory Work.

Most of these objects are readily attained by selecting for early treatment actions in which air plays a part. Historically, the discovery of oxygen by Priestley and Scheele, and the proof of its presence in, and responsibility for the chief properties of the air by Lavoisier, coming as it did at a time when chemistry was just crystallizing into a science, point to experiments on the action of air as particularly significant in an historical point of view. The study of oxygen, a gas, enables us rapidly, if we choose, to approach the theoretical portion of the science, and, at the same time, the familiar nature of its chemical effects makes them suitable for introductory work.

We assume then that some simple and more or less familiar facts, some of them probably connected with the action of the air, will be presented in the beginning to the pupil. Some of these may be examined by him personally in the laboratory; others may be shown him by the teacher. It is impossible for us here to describe in detail the method which the teacher will pursue in bringing out the significance of what is seen by calling attention to the detail of observation and leading the pupil to the interpretation of each detail. Illustrations of the method to be used will be found admirably given in Perkin's work already mentioned. Professor Richards gives some instructive examples in the Harvard pamphlet of requirements in chemistry.[1] Some laboratory manuals also develop the method of instruction with considerable fulness.

III. Earlier Generalizations of a Qualitative Nature.

First Characteristic of Chemical Change.

The first thing which the examination of several chemical changes reveals is that a total alteration in all the physical properties of the substance takes place. This is a feature requiring minute and careful instruction. In order that the pupil may adequately appreciate this characteristic, the nature of the various physical properties which are interesting to the chemist must be discussed more or less fully, and in each particular case the properties of the body or bodies before chemical change, and the new properties after chemical change, must be carefully and exactly enumerated. The pupil will not do this for himself, and without it his ideas must remain somewhat hazy. This is advised not because the *doctrinaire* treatment of this conventional sign of chemical change is particularly helpful, but because a keen appreciation of physical details is at the basis of all chemical work. The pupil must eventually learn to recognise materials by their physical properties, since by this alone can he study chemical change qualitatively.

Importance of Knowledge of Physics.

In many cases the study of physical properties is treated as a separate topic before any chemical change is introduced, and there is certainly justification for this course, even if the pupils have already studied the science of physics itself. There are many physical matters important to the chemist which are not treated in elementary

---

[1] *Requirements in Chemistry for Entrance to Harvard College.* Cambridge, published by the University, 1900.

physics.[1] Every teacher of chemistry knows the mistakes which the beginner makes when told to evaporate any substance in order to obtain crystals for examination. The pupil is utterly innocent of any knowledge of the conditions under which crystals are formed, and usually does not even recognise that the amorphous mass he obtains by violent boiling over a naked flame is not the required crystalline product. And this is only one example out of many which might be adduced to show that a knowledge of physical properties which remain unconsidered in school physics is an indispensable part of the equipment of the pupil in chemistry. The necessity of attention to this matter has already been emphasized (pp. 30–33 and 39), and its extreme importance alone justifies our recurrence to the subject.

This study of physical change leads to the familiar generalization[2] that every physical property is altered, and that the alteration is usually permanent. Most frequently the matter is made clearer by contrast with physical change. It is advisable also to cite familiar instances of each kind of change.

The study of the facts in connection with the first few experiments next reveals the nature of chemical change. Some material has come out of combination or gone into combination. In other words, the great change in physical properties is accompanied by a change in composition. If the experiments on which this conclusion is founded have been properly selected, they will incidentally lead to the classification of changes in composition into the three common kinds. The

Second Characteristic.

[1] In Tilden's *Teaching of Elementary Chemistry*, 8–11, and more especially in the work of Perkin and Lean which we have already (p. 56) mentioned, 22–125, will be found laboratory instructions covering a large number of experiments on those physical properties, familiarity with which is most important in chemical work.

[2] The two or three facts actually in the hands of the pupil do not, of course, strictly speaking, justify generalization. No generalization in a science deserves the name unless it is founded upon an immense range of facts. The teacher must therefore indicate the direction in which numerous other facts of the same kind lie, in such a way that the pupil readily appreciates their nature, and feels satisfied with the general principle deduced. The conscientious development of a single generalization might otherwise occupy the whole year.

first experiments in chemical change are discussed again in this connection.

The pupil's attention may next be drawn to the production or disappearance of heat in connection with some of his illustrations of chemical change. Special experiments may even be introduced to show that in like manner light and electricity may be consumed or produced in a similar way. This is not of course the place in which to discuss energy, but it furnishes a convenient opportunity at least for drawing the attention of the pupil to the fact that all chemical change is accompanied by energy change of some kind. Perhaps even the economic importance of this in connection with the steam-engine and the storage battery may be referred to. If he has already studied physics, the tendency to the dissipation of energy, which a chemical system, in common with any physical one, exhibits, may repay notice. In any case, none of the subjects touched at this stage can possibly be treated fully or become a section of the subject complete in itself. Usually, recurring to the same subject at intervals, and adding a little each time, will be more effective, when the question is an abstract one, than a complete treatment of it at any stage. The pupil becomes gradually accustomed to thought about the subject, and thus does not experience the difficulty and perhaps disgust with which a sudden presentation of abstract ideas may otherwise affect him.

Third Characteristic.

Aside from the three main features which we have mentioned, there are matters which may be described as minor, and which yet are exceedingly important and soon begin to obtrude themselves upon our notice. There is, for example, the necessity for contact in order that chemical action may occur. It is long before the pupil realizes that putting two materials in the same test-tube is not the equivalent of giving them every opportunity for interaction. If, for example, the experiment is to place powdered potassium iodide in a test-tube, add concentrated sulphuric acid and observe the result, one pupil will fulfill the directions to the letter, while his

Other Minor but Important Truths.

neighbour may use large crystals of the substance instead of powdering them. Thus while the former obtains a violent action in the cold, the latter may decide that practically nothing happens. It requires most persistent discussion to lead students to realize that, unless means is taken to permit complete access of every part of each substance to every part of the other, the best conditions for chemical change have not been fulfilled, and that, without thorough mixing, chemical action is as little to be expected as if the substances had been in different test-tubes instead of the same one. Another matter worthy of notice is the great increase in the speed with which a chemical change takes place when the temperature is even slightly elevated. This, together with the melting or other assistance to contact which heating affords, is the reason for its effect on chemical change. A third point, which for the present is of minor importance, is the fact that chemical changes are often carried out with incomparably greater ease by dissolving the substances in water, and that in most cases of this kind the water is not a factor in the change. It is only by noting matters like these, in the many various ways in which they affect chemical change, that the pupils' chemical intelligence can be slowly developed.

IV. Further Generalizations, of a Quantitative Character.

The basis for the introduction of the fundamental quantitative laws of chemistry may soon be reached. This may be found partly in the very first experiments, and partly in additional ones designed more specifically for the purpose. If it is desired, the systematic development of the subject-matter may begin at this point, or at all events immediately after the first of the following principles. This, following the historical order, will probably begin with a more formal study of oxygen and its relation to air. Or, as some writers prefer to arrange it, hydrogen may precede oxygen, and water may precede air. We shall not attempt to express any preference in regard to the particular time for introducing this treatment or the particular topic with which it shall begin. The matter is largely one depending on the taste of the teacher, and the arrangement of the book he uses.

The first of these generalizations arises naturally in answer to the question whether, in the changes which have been noticed, one body combines with another and alters the character of the latter without adding to the weight, or whether each substance takes its weight with it into combination. This being answered in the affirmative, the further question arises — whether this occurs absolutely without loss or gain, or takes place with some slight abatement or modification of weight. The fact, of course, is that, of all the physical properties of a substance, its weight is the only one which it is found to have carried with it through any number of chemical transformations.[1]

Fourth Characteristic.

It must be clearly explained to the pupil that this principle cannot be rigidly established without an immense number of experiments, and all of them would have to be of a more exact character than the technical skill of the beginner could furnish.

Closely associated with the question answered by the previous law, is that of whether, in producing the same compounds, constant proportions of the constituents are required. The answer is naturally in the affirmative.

Fifth Characteristic.

[1] Phrases like *the conservation of matter* or *of energy*, if used at all with beginners, should be defined carefully in strict harmony with their particular experience, or, if they have none, at least with conceptions which can most readily be pictured to the mind. The statements, for example, that the "sum total of each kind of matter," or "of all the energy" "in the universe" is constant are too remote from experimental examination to be seen to have any relation to ordinary experience. It may be remarked also that they are not in this form scientific statements, but metaphysical speculations. All that we can verify by experiment is the fact that in physical and chemical operations on a limited scale the matter and energy can all be accounted for, and we have no evidence that any is lost or gained. The more abstract mode of statement leads the pupil naturally to think that these laws are simply dogmas. Many of us, having received this false impression, have for a time wondered greatly what the origin of these dogmas was.

As before, these generalizations will be illustrated by reference to every-day experience on which they have a bearing. Generalization is not an end in itself. It is simply the clear formulation of a fact preparatory to its employment for illuminating our experience. Application in later work in chemistry occurs as a matter of course. It is important, however, that the employment should be as wide in range as possible. We are all familiar with the surprise with which the obviousness of an application or illustration strikes us after the relation has been pointed out by some one else. Yet it is chiefly an independent ability to apply what is known that distinguishes the scholar from the prig. The mastery of the generalizations of chemistry may constitute a part of learning in a narrow sense: to have digested them and become able to see their application to remoter facts within our knowledge is education. The possibilities and methods of application are discussed more fully under classroom instruction (chapter V., sections a and e, pp. 129 and 138).

Measurement of Proportions by Weight.

In connection with the discussion of the law of definite proportions, the question of the actual ratios by weight in some simple chemical compounds will naturally come up. The proportions in some of the actions already noticed should be given as illustrations, and the results expressed by percentage. If possible, the actual carrying out of a measurement should be shown. The union of a weighed amount of copper with oxygen, for instance, is a suitable experiment, for it requires no supervision. Other quantitative experiments which are available will be discussed in the chapter on the laboratory work.

Sixth Characteristic.

The principle of multiple proportions may fitly follow. As a classroom illustration, the reduction of cuprous and cupric oxides by hydrogen will be found easy, provided pure cuprous oxide can be obtained,[1] as failure to get good results is almost impossible.

The next generalization is that relating to reciprocal propor-

[1] I have found Kahlbaum's most satisfactory.

tions (law of combining weights). For its development a number of actual combining proportions and equivalent weights are required, and may be tabulated on the blackboard. The study of the numbers which a suitable series exhibits brings out a very remarkable fact about chemical combination. This may be stated as follows: If we take any element as basis, and any number as the value for the combining weight of that element, then the quantities of other elements which combine with this amount, or are equivalent to it in chemical combination, have this property, that complete combination of the elements with one another takes place when these quantities, or simple integral multiples of them, are employed, and no compounds are known whose composition is not in harmony with this rule.[1]

Seventh Characteristic.

This relation furnishes us with a set of combining weights, or rather, by varying choice of the basal element and value assigned to it, an indefinite number of such sets of weights. It may therefore be indicated at this point that convenience de-

[1] It is one of the most serious defects of many elementary text-books that they do not formulate this principle in terms of its experimental basis. It seems sometimes to be left entirely out, and its consequences creep in unawares under the cloud of dust raised by the atomic theory, or appear in the use of equations without any attempt at justification of the prodigious logical hiatus which this involves.

The experimental fact, stated in one way, is as follows: We take a definite quantity of an element A, and ascertain the quantity of an element B which unites with it. Then we measure the quantity of C which unites with this quantity of B; then that of D which unites with this amount of C, and so forth. We thus obtain a series of numerical results (equivalents) such that each quantity in the series is that which unites with the neighbouring quantities of adjacent elements on each side of it. Now we discover that the stated quantities of remoter elements are also such as enter into combination, either as they stand or with the use of the principle of small integral multiples. It is this fact which enables us to assign individual combining (atomic) weights to the elements. Without it, chemical proportions would be a waste of unrelated percentage compositions, and our much cherished formulæ and equations would have no existence.

The matter is explained with exceptional clearness in Young's *Elementary Principles of Chemistry*, 23–26, and 242–243. See also Vaughan Cornish, *Practical Proofs of Chemical Laws*, chapters I. and IV.

mands that some particular set shall be preferred. It is clear that numbers less than the hydrogen equivalents will be inconvenient, as they must either make hydrogen itself less than unity or introduce unnecessary multiples whenever they are used. The selected combining weights (atomic weights) may be given, and will be seen to be frequently small multiples of the hydrogen equivalents.[1] The basis of selection cannot be further explained without the use of the consequences of Avogadro's hypothesis.

This point in the development of the principles forms a convenient halting place, and we shall not pursue the subject further at present. The results of the work we have outlined suffice, if the teacher so desires, to enable him logically to introduce symbols and equations. At this point, or a little later, if he sees fit, he may also present the explanation of the last three generalizations which the atomic theory furnishes. The discussion of the relations of symbols (p. 77) and of the atomic theory (p. 154) to introductory work will be taken up later.

Avogadro's Hypothesis.

It will be seen that the chief theoretical subjects affecting the quantitative description of chemical change which still remain for consideration are: Avogadro's hypothesis and its application through measurement of the density of gases to the determination of molecular weights, the final adjustment of combining (atomic) weights, and the explanation of valency. It might be noted at this point that many teachers do not favour a *complete* discussion of these subjects in the secondary school. They are undoubtedly difficult, and must necessarily occupy a great deal of time, and, when all is said and done, the pupils are little likely long to retain much of the intricate reasoning which is inseparable from their discussion. It is true that, like any other part of the science, the study of this aspect of it must furnish admirable discipline, but it is a

---

[1] Throughout, oxygen equivalents ($O = 8$ and $H = 1.0076$) may be used just as easily as hydrogen equivalents ($H = 1$ and $O = 7.94$), and they have the advantage of leading directly to the standard atomic weights ($O = 16.00$).

question whether even in this point of view a more economical use of the time may not be made by substituting other and simpler chemical topics. These particular things are not likely to find application in every-day life, even if they are retained, and, in the less usual case of the pupil who afterwards attends a university or technical school, this subject will in any case have to be dealt with afresh. It is on account of these facts that I am inclined to justify the less rigid treatment of the matter of combining weights, in order that apart from Avogadro's hypothesis we may have a reasonable basis for the use of equations.

V. The Relation of the Quantitative Laws to Formulæ and Equations.

One of the chief criticisms of the teaching of chemistry at the present day is that much of it fails to make clear the place of the balance in chemical work and the relation of the results of measurement to the plan chemists have adopted of expressing these results, namely, by the use of the combining or atomic weight as the unit of quantity for each element. I am not, for the moment, referring to the much debated question whether the pupils can or should do quantitative work. It is the unassailably fundamental character of the quantitative data and their interpretation that I would emphasize. It is this that has made chemistry an exact science. Thus, even if he has no balance at all, or no inclination to use it, the teacher is still compelled to reach the core of the science, if he reaches it at all, by explaining, in one or two actions at least, how one could set about measuring the quantities concerned.

Expression of Quantities by Formulæ.

When the time comes for expressing these measurements in the form of symbols and equations the pupil must be shown clearly how the translation into the conventional chemical formulæ is effected. It may seem to some readers a strange statement to make, but I believe that many will bear me out, when I say, that, although the modern works have included the stage of measurement, there are few elementary text-books in which any attempt is made to furnish the links between experiment and equation. I know hardly any that I could put into the hands of an intelligent person for study, with the least confidence that this connection would be understood.[1] It is to be feared that the number of teachers who furnish this link must be limited, for the books must by all means represent the average, if not the best teaching in the country.

A concrete illustration will make most clear what is meant. Suppose the teacher deems that the time has come for the use of formulæ and equations to begin, and that the introduction has been conceived somewhat in the spirit of the preceding section. To be specific, suppose that he decides to do this in connection with the study of oxygen. Let us further suppose that sulphur is the first body whose union with oxygen is observed. After qualitative observation the question of quantity arises. It is necessary to ascertain, or assume as known, the weights of two of the three bodies, sulphur, oxygen, and sulphur dioxide. The third can then be inferred. The simplest experimental method is that which weighs the sulphur and burns it in excess of oxygen, and catches and weighs the sulphur dioxide. The apparatus and general procedure must be sketched and described or shown.[2] The result

An Illustration.

---

[1] The explanation is admirable in Reynolds, *ibid.*, 69–72; it is clear, but too long postponed in Torrey, *ibid.*, 315–316; in Young, *ibid.*, 75–76, it is satisfactory.

[2] Newth, *Elementary Inorganic Chemistry* (Longmans, Green & Co.), p. 108, describes this experiment. It will do very well for description, but I do not advise its performance, as it requires careful watching, and I have found that the boat, or glass of the tube, often acts catalytically and sulphur trioxide is formed in such quantities that the weight of the sulphur dioxide, and therefore of the oxygen, comes out much too large.

Perhaps the best experiment for illustrating the making of a formula is the solution of a weighed piece of iron wire in nitric acid, and the evaporation and ignition of the residue ($Fe_2 O_3$). The indirect nature of the oxidation is unfortunate. But the oxides which are formed easily by direct union, like those of copper and magnesium, do not afford an example of the use of multiples of the combining weights, while there

of the experiment leads to the conclusion that the proportion by weight of sulphur to oxygen is 50 : 50 in a hundred parts, or 1 : 1, almost exactly.

Now chemists express this result in a system in which the combining (atomic) weight of each element is the unit. There fore, what we desire next is to know the value of x, the number of combining weights of sulphur, and y, the number of combining weights of oxygen in the equation:

$$x \times \text{comb. wt. of sulph.} : y \times \text{comb. wt. of oxygen} = 1 : 1.$$

The combining weights must be known or the operation stops here, and the equation cannot be reached until they are known. We state them to be 32 and 16 respectively. The problem then is to find the simplest values of x and y in the equation $x \times 32 : y \times 16 = 1 : 1$. If the combining weights have been successfully chosen, x and y must be rational numbers, and will usually be small numbers. This is the property of chemical combination mentioned in last section (p. 75), in consequence of which alone we possess interchangeable combining weights of any kind at all. Here evidently $x : y = 1 : 2$. Now the symbol S expresses 32 parts of sulphur,[1] the combining weight, and

---

are objections on the score of experimental difficulty to the use of the oxides of carbon and phosphorus.

[1] This statement of the meaning of the equation harmonizes with the mode of approach from the experimental side which we have pursued. In this point of view, S may not be used as a contraction for the name of the body. Nor does it mean an atom of sulphur, since atoms are not perceived in experiment.

The symbol represents, not the yellow, light solid which is indicated by the word sulphur, but the part of the mass of the compound (in this case, sulphur dioxide) which was originally sulphur, but now shows none of its properties. Some chemists distinguish between the free body, or "simple" (as opposed to compound), and the element. The latter is the same material in its combined and unrecognizable form. The atomic theory, in explaining the quantitative laws of combination by supposing each constituent to be done up in little pieces of uniform weight, incidentally leads us to suppose also that the pieces remain intact after combination. It suggests that the pieces are stuck together without losing their individuality. So the symbol SO_2 shows them to us side by side. But precisely because the theory and the atomic ideas conthe symbol O stands for 16 parts of oxygen, its combining weight. Therefore the composition of sulphur dioxide, in terms of the chemical units of quantity is $1 \times S : 2 \times O$, which is equivalent arithmetically to the proportion by weight, 1 : 1, found in the experiment. The formula is thus SO_2. Since one combining weight of sulphur and two of oxygen are required to form this, the equation is $S + O_2 = SO_2$.[1]

Every step in this process is easily followed, and the equation is seen to rest directly on experiment, as it should do in a science which claims to be experimental. The only point of possible obscurity is in the justification of the atomic weights. If Avogadro's hypothesis has not been given, that difficulty cannot be fully met. The pupil has seen, however, how a set of combining weights can be established, and may, without confusion, be asked to leave in suspense the question of why these particular values (such as 32 for sulphur rather than 32/2 or 32/5) have been finally chosen. This whole proceeding must be repeated with each succeeding equation for some little time, until it is thoroughly familiar. If the teacher can actually perform one

---

nected with the formula prejudice us in favour of the view that the sulphur persists in the compound in this way, we are apt to forget that the theory makes no attempt to explain why the product is a wholly new species of matter. It simply ignores the fact that a body with a powerful odour and the other familiar physical and chemical properties of sulphur dioxide has appeared, and that the characteristic properties of the two bodies from which it was made have been submerged (*cf.* Perkin and Lean, *ibid.*, 322). In other words, we are put in the risky position of trying to think that the bodies are both still there and both gone at the same time. The qualitative facts are better explained by supposing complete change of the sulphur and oxygen and production of the compound out of the material (*cf.* pp. 156, 158). Thus, in the formula of the compound SO_2, at least, the material denoted by S is not free sulphur (the "simple"), and the symbol and word are not interchangeable,—at all events, not until the atomic theory has been given, and, strictly speaking, not even then.

[1] The general solution of this problem consists in taking the weights of the constituents found in any sample of a compound, dividing each such weight by the combining weight of the corresponding element, and finding the whole numbers which stand in the same ratio as the quotients.

experiment before the class, the impression will be incomparably more definite and lasting. If it is possible to include such an experiment in the laboratory work of the pupils, the effect will be still better. But my point is that the equation can never be understood unless the quantitative measurement, whether by description, demonstration, or individual performance makes relatively little difference, is brought to the notice of the pupil, its numerical result seen, and its translation into the form of an equation exhibited. Without some such explanation, the equation is bound to be a mysterious thing, and must remain utterly unconnected by any visible link with the chemistry of the lecture-table and the laboratory.

Symbols used to Represent Weights, not Atoms.

It will be seen that we have treated the symbol and the equation as if they represented the materials, not by atoms and molecules but by weight. It is frequently assumed that the reverse of this is the correct view. Yet the fact is that the equation is seldom used in the latter sense, and there is no impropriety in treating the former as its primary signification. Chemistry is primarily experimental.[1] To illustrate. Marshall (JOUR. CHEM. SOC., Lond. LIX., 771) found a white crystalline body in a cell, originally filled with a solution of potassium bisulphate, through which a current of electricity had been flowing for a long time.

[1] The equation may therefore be used solely as a record of quantitative data until the atomic theory has been introduced. After that has occurred, the explanation that the chemical unit quantities of each element are likened in this theory to atoms will naturally be given. When Avogadro's hypothesis has finally defined the chemical molecule, the equation comes up once more. This time it has to be changed to correspond with the new ideas, if the items it contains are to be complete molecules, and the equation is to represent the change as if it took place in the physical minimum of materials. Where we can measure the molecular weights,—as in the case of volatile substances,—the formulæ will now be adjusted so as to represent molecules (as $O_2+2H_2 \rightarrow 2H_2O$) and the equation will embody the change *in petto* physically, as well as in proportions by weight chemically. For practical purposes this form of equation is preferable only when gases are concerned. With bodies which, under ordinary conditions, are not gases (as P_4), it is a needless complication.

He did not for a moment attempt to ascertain the nature of the body, and the action by which it had been formed, by trying to see how the molecules of the bisulphate were affected by the current and what atoms were contained in the molecules of the new substance. If the chemist did not think in a more concrete way in his work than he seems often to do in his teaching, his mind would be so befogged with theory that he would never accomplish anything. Marshall simply analysed the body, found that the elements potassium, sulphur, and oxygen were present in it, and determined the percentage of each. A process of reasoning essentially the same as that in our illustration gave the formula KSO_4. Comparison with the bisulphate ($KHSO_4$) showed that hydrogen had been eliminated, and, as this gas was liberated during the electrolysis, the simplest equation was evidently: $KHSO_4 = KSO_4 + H$. This was the first isolation of a persulphate.[1]

Naturally only a mere fraction of the actions studied in an elementary course can be treated quantitatively. But when, without exact measurement, other equations are constructed by the help of such data as the pupil's experiments afford, or are even taken from the book, the learner will still realize that the process described above was carried out by some one, and the origin and meaning of the equation, once grasped, will never again become obscure.

Equation Writing.

'Equation writing,' in the sense in which the phrase is commonly used, will be a necessary exercise in connection with the making of the record of each experiment in the note-book, but, after rational explanation like the above, it will be largely a genuine exercise in chemical thought, based on inferences from observation. Without any such explanation, it is likely to be a combination of copying and illogi-

---

[1] Subsequent study showed that persulphuric acid was a dibasic acid, and that therefore the formula of the salt was $K_2S_2O_8$. The above process could not go beyond correctly expressing the proportions by weight in terms of multiples of the atomic weights. Entirely different means were required to furnish this improvement.

cal collocation of letters, accompanied by vague gropings after some rule. If 'rules' are supplied, then the last chance of any chemical benefit surviving is removed and some puzzle like the once popular "fifteen puzzle" would furnish more exercise for the intelligence and just as much chemistry.

If the exercise is merely arithmetical like this:

"Complete the following:

$$CaCO_3 - CO_2 = ?$$
$$2\ KNO_2 = N_2O_3 + ?$$
$$Pb + HNO_3 = Pb\ (NO_3)_2 + H_2O + NO,"$$

a little of it may be safe and useful. Even students in universities can seldom count the numbers of atomic weights of each element correctly. But too much of this sort of thing suggests to the pupil that a chemical change can be predicted safely from manipulation of the formulæ of the factors entering into it, and withdraws the attention from careful reasoning from observation.

This is especially likely when, as in the book from which the above is taken, a large number of such truncated equations are given early in the course, before the actions with which they deal have been studied. We are not surprised a few pages further on to read: "HCl is the formula of an acid because it[1] consists of hydrogen united with a non-metallic element," followed by an exercise in selecting "from these symbols" "the ones which stand for acids:" "$H_2C_2O_4$, $CaCO_3$, NaOH, H_2S." The reasons for the decision are to be given. How unpleasant it would have been if the printer's devil had maliciously added H_6C_2O, H_8NO, and $P(OH)_3$ to the list!

The Time for Introducing Equations.

In view of the fact that equations, symbols, formulæ, etc., are not parts of chemistry, but of our mode of recording chemical facts, it seems desirable to plant the facts and their importance securely in the mind of the pupil before these conventions are considered. When they are

[1] This reminds, one irresistibly of a pupil's answer quoted by Tilden: "Metals differ from non-metals, both by ending in *um* and having a metallic lustre."

given they should be clearly distinguished from the facts of the science and explained as a kind of abbreviated language for expressing the facts.

Since they are quantitative expressions, they cannot logically appear before the method of measurement has been explained. If the problem were that of explaining to a visitor from Mars the system of money and exchange used on this planet, we should not first invite him to memorize part of the stock exchange and commercial reports of a newspaper. He would have to become familiar with the materials themselves, and understand the object and machinery of exchange, before he could intelligently handle the conventions by which we have come to chronicle them. The introduction of equations soon after the quantitative laws have been explained and illustrated is desirable, because their use gives much greater precision to the pupil's conception of each chemical change. On account of the abuse to which they are subject, their postponement to a very late stage, or even omission altogether, has been seriously advised by many teachers. This, however, is surely an extreme view. Why propose excision of a valuable organ when the cure of the disease lies in the hands of the teacher?

CHAPTER IV

INSTRUCTION IN THE LABORATORY

We have already, either by implication or directly, touched upon many of the important aspects of instruction in the laboratory. This was unavoidable, since the purposes and results of chemical instruction are so closely bound up with the practical side. We must now discuss the subject in a more systematic manner. The value of laboratory work offers a convenient title under which we may endeavour, through the discussion of the benefits it may confer, to set forth the considerations which affect the attitude of the teacher, and are to be used in moulding that of the learner, towards the science. We may then consider in succession, the laboratory directions, the attitude of discoverer or verifier which the pupil may be induced to adopt, the importance of technique, the question of quantitative experiments, and the note-book.

In all this, of course, we assume that the teacher is provided with a laboratory of some kind, and with equipment more or less adequate. The construction and furnishing of a laboratory will be dealt with in a chapter by itself. The importance of its right employment in teaching chemistry will not require separate treatment, since it will emerge very distinctly from the results of our discussion of the topics immediately following.

Laboratory Instruction as it is.

The laboratories, in many cases magnificent, with which new school buildings are nowadays usually provided, show that the necessity for having them has been recognised, at least by the architects. One would fain think that their importance is equally appreciated by superintendents and principals. The statement made by the Committee of Nine, however, although startling to those who have looked into one or two of the best schools, and have not taken the average of the secondary schools of a whole state into consideration, is, it may be feared, not without justification. They say:[1] "While the laboratory method is almost universally approved by the science teachers, the text-book method prevails in the schools, to such an extent that laboratory work is incidental, inefficient, and in many cases excluded altogether." In their preliminary report, which deals with this phase of the subject more fully, they say:[2] "It is true that attempts at laboratory work in one or two subjects are reported by the schools of the State of New York almost without exception, but the complaint is made that the laboratory study must be limited, desultory, and subordinate to the study of books with classroom exhibitions." They attribute this condition in part to lack of recognition of "the great labour involved in the conduct of good laboratory work," and the fact that "the reputation of the teacher and the standing and financial support of the school are affected by the results of examinations," while "the results of laboratory study cannot be tested by the current methods of examination." As this committee included representatives from the high schools, as well as the normal schools and colleges of New York State, and as this opinion was expressed after prolonged and minute study of the actual condition in the schools of a state which is certainly not below the average educationally, it must be regarded as a serious criticism of the present teaching in the whole country. Only the constant efforts of the teachers of science, to awaken the minds of other school officers and of the public at large to the indispensability of reasonable expenditures for scientific equipments, and reasonable consideration of the points made by the committee, can secure the gradual remedy of this state of affairs. No special benefit is to be expected from the teaching of science until a far-reaching change has been brought about.

---

[1] University of the State of New York, *High School Bulletin No.* 7, 706.

[2] JOURNAL OF PEDAGOGY, Vol. XI. (1898), 119.

I. The Value of Laboratory Work for General Education.

a. *For Teaching Knowledge-making by Observation and Induction:*—Observation in chemistry implies something much more complex and difficult than we sometimes appreciate. In its simplest terms it may consist in noticing the colour of a precipitate, or stating whether bubbles of gas do or do not appear, or perhaps in describing the form of a crystal. This demands what one writer has described as "ocular accuracy." The process is one of the mind, although the phrase suggests that the eye as a physical instrument is mainly concerned. In many experiments, however, the use of experience and reasoning in observation so greatly predominates, that the part which the eye or the sense of touch plays becomes relatively inconspicuous. We read of observation consisting in the "training of the senses." The phrase is vague. It should be remembered that the reactions of our sense organs are scarcely affected by practice. Galton has shown that sailors' eyes, instead of being more efficient physically than other peoples', are really less sensitive than the average. It is the ability of the seaman to interpret what he sees in the light of experience that makes him a better observer of some things than the landsman. A boy may see ten times as much as a man, yet the man will learn ten times more from what he sees than the boy. Applying this to the matter in hand, we see that the training of a pupil in observation consists really in storing his mind with suitable experience, all thoroughly classified and digested. Ability to observe chemical phenomena is an attribute of the chemist, and teaching observation consists really in teaching chemistry.

What is implied in Observation.

b. *For Teaching Knowledge-making by the Study of Natural Objects and Phenomena:*—As we have seen (pp. 10 and 49), training in observation in the fullest sense of the term may be obtained from the study of languages and other book work. The man who is trained, however, in this direction only, may remain bookish and un-

Practical Value of Laboratory Experience.

practical. The knowledge by which we live is not furnished with an index, nor is it arranged alphabetically. It is thrown at us much like the experience of the chemist, and, as a school of education and a sphere of activity, the world is more like a laboratory than a library. The experience in chemistry quickly shows the fallacies into which we continually fall, and from which experiment and renewed observation alone can rescue us. We quickly learn that the operation of thinking clearly and keeping our ideas in touch with facts is not a natural attribute of the untrained mind. In studying chemistry in the laboratory we acquire the habit of applying to concrete things the methods of observation, of induction, and of testing every hypothesis by reference to facts, which are indispensable to clear thought about such matters. The application of the method is a quality of the scientific mind, whether that mind is employed in business or in study. As Professor Remsen[1] says: "By a scientific mind is meant one that tends to deal with questions objectively, to judge things on their merits, and that does not tend to prejudge every question by the aid of ideas formed independently of the things themselves."

Illustration of Faulty Induction.

To illustrate:[2] When ammonium chloride is heated in a test-tube and litmus paper shows the presence of ammonia at the mouth of the tube, the student instantly says that ammonia has been given off, and thinks of the remaining solid as containing the hydrogen chloride. Presently the test paper shows the arrival of this acid, and he is reminded that the other product is a gas. If he is now asked why the ammonia appeared first, he will invariably say that it must have been formed before the hydrogen chloride! It usually comes as a surprise when you lead him to see that in the decomposition of a substance

---

[1] Address on "The Chemical Laboratory," delivered in connection with the opening of the Kent Chemical Laboratory of the University of Chicago. NATURE, XLIX. (1894), 531.

[2] An excellent discussion of this, with many historical illustrations, is given by Tilden, *Hints on the Teaching of Elementary Chemistry*. London and New York, Longmans, Green & Co. 1895. Pp. 1–11.

into two gaseous molecules, the products cannot be formed otherwise than simultaneously. The same pupil would not have made so grotesque a mistake in reasoning in geometry or in translation from French. It is the reasoning about material objects and phenomena which is difficult to him because it is unfamiliar.

c. *For Teaching Caution and Mental Rectitude*:—The illustration just given points further to the continual discipline which the pupil must receive in the necessity for caution in forming conclusions. No work so much as that in chemistry impresses one with the necessity for distrusting preconceived notions, or furnishes a better preparation for tenaciously employing this principle as one of the best guides in all the actions of life. The line between the minimum inference which the facts actually justify, and the more extensive one which we are continually tempted to draw, is often so easily passed that the most varied experience in searching for it and remaining on the safe side can never make the process too familiar. Take, for example, the case of the Bunsen burner. When the pupil, after he has observed the difference which the position of the ring at the bottom of the burner makes, is asked what the openings have to do with the matter, he will invariably say that it is the admission of oxygen which causes this difference. This statement may even be found in many books, in spite of the fact that nitrogen and other gases which contain no oxygen have the same effect.[1] The higher temperature of the Bunsen flame is sufficiently explained by its smaller size. The liberation of free carbon in luminous flames is another rock of offence. It seemed to be accounted for by the theory that the hydrogen of the hydrocarbons burned more easily, until this was shown conclusively to be the exact contrary of the fact by Smithells.[2]

Need of Care in Drawing Conclusions.

The custom of careful scrutiny of hypotheses and their con-

[1] See Newth's, *Inorganic Chemistry*. London and New York, Longmans, Green & Co. 1894. Pp. 291–306, particularly 304.

[2] NATURE, XLIV (1893), 86. For further references, see p. 215.

tinual probation before the court of experiment begets a habit of mind which finally finds delight in the search for exact knowledge and correct opinions for their own sake. In these days of exaggeration and superficiality the influence of the tendency of laboratory work to the fostering of mental rectitude cannot be prized too highly.

d. *Other Benefits of a General Nature:*— Laboratory work is undoubtedly of value in that cultivation of the mind which is expressed by care and neatness in mechanical matters, and in dexterity in the manipulation of materials. This training has undoubtedly a broader significance, beyond the operations and objects peculiar to chemical work. Perhaps, as a substitute for, or supplement to manual training,[1] it may be said to have some value in a partial way.

In all instruction the personality of the teacher is held to be a factor of not less importance than the nature of the subject taught. The close personal contact which laboratory instruction secures between the pupil and teacher, and the consequent greater opportunity which his personality has to impress itself upon the pupils, is not one of the least of the benefits we are discussing. There are others that might be mentioned; several, such as the liberation from the bondage of authority (pp. 10 and 51), have already been discussed in other connections. The examples we have given (in b and c above), if carefully thought out, will furnish a clearer insight into the value of laboratory work than any mere enumeration of ours could do.

II. Value of Laboratory Work for Instruction in Chemistry.

a. *For giving First-hand Knowledge:* —The study of chemistry, or any other body of knowledge, must be carried out by direct encounter with the material of the science itself. The study of what some one else has said or thought about the subject is an interesting, but entirely different exercise. We do not study Latin by reading an English translation, or an essay

[1] W. E. Bennett, Manual Training of Chemistry. University of the State of New York, *High School Bulletin No. 13*, 926.

on the author's work. Every one understands that the study of Latin means the study of the text itself. So the term "study of chemistry" can be properly applied to nothing but laboratory study of the subject. An author's explanations and verbal statements are but a feeble and exceedingly partial substitute for the facts themselves. Really to know what the facts of chemistry are, they must be seen and handled directly. The books are not chemistry, but literature, and, as some one has said, they are mostly poor literature at that.

The Study arranged round the Laboratory Work.

In order that there may be no question of the pre-eminence of practical experience, the course should be arranged round the laboratory work, and the latter should carry the thread of the subject. Classroom work and other exercises should be adjusted to this and used as supplements. This of course presupposes the existence of certain qualities in the chosen laboratory outline which shall fit it for furnishing the backbone and carrying the burden of the work. It is evident that if the relation of the other parts to the laboratory work is not that which we have suggested, if, for example, the recitations from a text-book form the only continuous and logical feature of the course, the attitude of the student towards the laboratory work will be entirely false. If the text-book is taken as the basis, and the impression is given that experiments are thrown in like the engravings and autograph letters in what bibliophiles call extra illustration of some book, they are bound to suggest mere ornamentation of some pre-eminently worthy nucleus, and the whole anatomy of chemical instruction must be deformed.

Psychology of Laboratory Study.

b. *For holding Interest and Attention:*—The whole psychology of laboratory work forms an interesting study in itself. Without attempting to treat the subject fully, we may draw attention to one fact at least which contributes to its value as a means of instruction. During the performance of an experiment, unless it is an exceptionally tedious one, it is almost impossible for the interest of the pupil to be withdrawn, or for his attention to flag. The operations being performed, the changes being watched, and the legitimate curiosity in regard to what will happen next, keep the whole matter constantly in the centre of the pupil's field of consciousness, and effectually prevent mind wandering. The strain on the pupil's powers of voluntary attention, which book work brings with it, is thus avoided in a large part of the time devoted to the study of chemistry. His thought about the subject also, with the activity continually prompted by this thought, satisfies the psychological demand for reaction as the necessary correlative of reception.[1]

Professor Dewey has pointed out another advantage which the laboratory possesses over the book, inasmuch as the performance of an experiment entirely diverts the attention of the pupil from the thought that he is studying, and fixes it completely on that which is being studied. In other ways of learning, the thought that he is studying is continually in danger of approaching a focal position in the field of consciousness, and relegating the object of study to a less central position, and even occasionally banishing it altogether.

c. *For Securing Clear and Pregnant Expression:*—In most studies we begin with the expression of the fact, and seek by study of the statement to reach the fact itself. In practical science, we encounter the fact first, and, having the fact clearly in mind, proceed to find a suitable expression for it. The former process is subject to misunderstandings which are only too familiar. Even if the language used is fortunately chosen, our personal equation,

The Order from Fact to Expression and not *Vice Versa.*

[1] *Cf.* James, *Talks to Teachers*, Chapter V. The pyschology of laboratory work has been admirably discussed by Newell in a paper on "More Profitable High School Chemistry." SCHOOL REVIEW, IX. (1901), 286. This article contains an admirable application to chemistry of the general principles discussed in Professor James' book. The criticisms of actual features of chemical instruction in the light of psychological principles will be found not only interesting but of practical value. It should be added that the *Talks to Teachers* form the clearest presentation of the application of psychology to teaching in existence, and in case it is not familiar to the reader its study cannot be urged too strongly.

resulting from the associations we have formed with the words, may result in more or less distortion when we seek to grasp the meaning. We are all familiar with the game of "rumour" in which the final result bears scarcely any recognisable resemblance to the original. Much instruction is of this kind. The teacher takes the fact he intends to present from a statement in a book. It went through several stages even before it reached his eye. But, leaving this out of account, we have first his conception of its meaning, then the expression which he gives it in conveying this conception to his class, then the interpretation they put upon his statement, and finally the effort they in turn make to reproduce it in their own language. The steps in this process are more than sufficient amply to explain the ludicrous misapprehensions which so often arise. In the laboratory the pupil encounters the fact directly, without the intermediate steps which involve the teacher, although the latter is of course concerned in assisting in the thorough exploration of the fact, and so the pupil is able to express the fact with much less risk of falsification.

Not only, however, are direct apprehension and clear expression of the facts of the science thus the privilege of the pupil, but the statement means much more to him after this process than it could have done if it had been furnished by the teacher or the book. It is not an assemblage of words or a dead phrase, but a statement bristling with reminiscence and significance. When we consider the limitations of language as a mode of expressing any idea with absolute precision and completeness, and at the same time without including too much, the advantage of this thorough grasp of the idea which has preceded the expression, and must forever accompany its use, will not require further justification.

An Illustration.

Let us illustrate by referring to one of the commonest forms in which the differences between related substances are expressed. The statement that chlorine is more active than bromine, and bromine than iodine, is a lifeless platitude to one who has no vivid experience with these substances to accompany it and give it meaning. After the elements have been handled, however, and comparison of their activity has been made by studying the action of chlorine and bromine on salts of bromine and iodine, for example, or by comparing the actions of sulphuric acid upon chlorides, bromides and iodides, or by heating the hydrogen compounds of the three elements, the word active acquires a definite experimental significance, and the whole phrase becomes pregnant with information.

III. The Laboratory Directions.

Since it is impossible for the teacher continually to supervise every motion and thought of the pupil, his place during the greater part of the work must be taken by printed laboratory directions. On the completeness and adequacy of these directions must depend to a large extent the realization of the purposes just discussed. If, for example, the instructions confine themselves to the barest statement of what materials shall be brought together, the pupil's experience will be utterly insufficient to furnish him with a conception of how best to perform the operation, of what to look for and when to look for it, and of the relations of the things he sees to one another and to his previous experience. The existing laboratory manuals show all sorts of directions, from the most meagre to the over-elaborate. It is, at all events, necessary that the teacher should carefully consider the directions in the book he uses, and adapt himself to them by preliminary discussion of the experiments. If need be he must give supplementary directions written on a blackboard, or perhaps substitute a more appropriate form in mimeograph sheets. The difficulty which this problem presents will be seen when we consider all the demands which may fairly be made on a good outline.

a. *Laboratory Directions Should be Coherent:*—The chief fault of laboratory study is its tendency to resolve itself into a series of isolated, and therefore mechanical, proceedings. The phrase we so commonly hear, that a pupil has performed fifty or a hundred experiments, suggests that this disintegration may by some be considered a merit rather than otherwise. Fifty experiments may contain the material for the development of a knowledge of the typical principles of chemistry, just as fifty sleight of hand tricks may contain the basis for the study of the psychology of illusion, but in both cases there must be a great distance to be covered before the results of the separate mechanical proceedings have been organized into a knowledge of either science. Evidently a long step can be taken in the right direction by arranging the operations in groups, with an idea running through each group, so that it shall constitute a study of some element or compound, or of the material for the development of some generalization. An example will show how a series of experiments, by proper grouping, may be converted into a systematic study.

Disadvantage of Incoherence.

Suppose that chlorine is the subject. The first question has to do with the ways in which it may be prepared. It may not be possible conveniently to illustrate all the distinct methods in the laboratory. The electrolysis of solutions of chlorides, for example, may be reserved for the demonstration. But the common general method, consisting in the oxidation of some chloride, will naturally be given. It is as easy for the pupil to take half a dozen test-tubes, and place in them various substances, like potassium chlorate, minium, barium peroxide, potassium dichromate, etc., and to treat each with hydrochloric acid, as to perform the same experiment with one of them. The view of the nature of the action will be broadened by further comparing the effect of litharge with that of minium, and of some metallic chloride and sulphuric or phosphoric acid with that of hydrochloric acid. The results naturally lead later to an instructive discussion of the meaning and mechanism of oxidation in this case.

Illustration, Chlorine.

Following this will naturally come the main preparation of a quantity of chlorine for the examination of its properties and the performance of some of the usual experiments with it. It may be noted that the union with phosphorus, antimony, and other elements are not distinct properties, but illustrations of one property, namely, the great activity it exhibits in uniting with various elements. Other distinct properties are its tendency to act upon water, forming a small amount of hypochlorous acid ($H_2O + Cl_2 \rightleftarrows HCl + HClO$. Note that the action of light on the solution is due to decomposition of the hypochlorous acid and is not a property of chlorine), its tendency to replace hydrogen in organic compounds, etc.

It will be noted that experiments giving negative results, like the action of litharge above, are useful rather than objectionable. They are, indeed, necessary in order that a basis for comparison may be furnished.

Illustration, Water of Crystallization.

In a similar way the study of hydrates, commonly spoken of as substances containing 'water of crystallization,' requires a number of closely related experiments in order that a basis for really understanding the subject may be furnished. We have to note, first, that a body like blue vitrol can be decomposed by heating and the product has entirely new properties; second, that this anhydrous material combines with water and the mixture furnishes crystals like the original ones; third, that the proportion used in combination can be expressed ultimately in terms of combining weights, proximately in terms of the formula weights of water and the salt, and is therefore genuine chemical combination; fourth, that the same substance may have crystalline form, of a different kind however, without containing water (it may be crystallized from concentrated sulphuric acid); finally, a number of crystalline substances may be examined by heating in order to ascertain which are and which are not hydrates in the common form in which they are sold.

There is no objection to the numbering of experiments. This is indeed an advantage to the teacher when examining the note-books. The point we have tried to make is, that the work must be grouped, and the groups must be coherent, in order that the results may be such that they furnish material for comparison, discrimination, and the arrangement in logical relation of the observed facts. When the results are assembled in such a fashion, the material is ripe for generalization. This final step will not usually be taken by the pupil, even if he is invited to take it. The review of the work in the quiz will be needed to bring out the relations more clearly, and in this exercise therefore the first development of generalizations will usually occur.

b. *Main Points in Regard to the Directions for each Experiment:*—The main features which are required to constitute proper direction in each experiment, and which must be furnished either by the manual, the teacher, or the head work of the pupil, may be summed up very briefly.

First, the object of the experiment must be definitely stated, or at least clearly implied in the title. Except in the case of verification of a law, however, it is obvious that the result of the experiment should be carefully concealed.

Directions in regard to Manipulation.

Second, the apparatus must be lucidly described, and if possible illustrated, in order that it may be readily constructed without loss of time; the object of the various parts should be mentioned in case any of them are new, or are not likely to be understood at once.

Third, a minute and practical description of the materials must be given. The quantity should be stated precisely, to avoid the tendency which the pupil generally has at the start to use four or five times too much; he should be shown that this results not simply in waste of material, but, what is much more important, in great waste of time. If solutions are concerned, the concentration to be used should be given: it will be noted that in general chemistry, unlike qualitative analysis, one strength will not serve for all experiments. Inasmuch as the state of many materials differs in different samples, the outline should specify whether an anhydrous or crystallized variety is to be used, whether the zinc is to be common granulated or chemically pure, or in the form of zinc dust, and whether lumps of the substance will serve the purpose, or whether it must be powdered. The great difference in the results depending on these things may be pointed out when opportunity offers (pp. 102, 131).

Fourth, the handling of the material and apparatus must be made clear. In simple cases like precipitation, for example, the very gradual addition of the reagent, accompanied by continual agitation, must be directed to avoid confusion. The curious layers which otherwise arise may else become the subject of observation and divert the attention into fruitless channels. Perhaps a special exercise on this is advisable. When the experiment is elaborate, minute directions are even more necessary.

The three last points are concerned with the peculiarity of chemical work, that the subject of observation has to be created by the pupil, and the lesson it may teach cannot be reached unless care is taken that the data, consisting in the phenomena observed, shall be specific and identical in every repetition of each experiment.

Directions concerning Observation and Inference.

Fifth, the point at which a pertinent observation may be made should be indicated by an interrogation mark, or in some other way. In one experiment the pupil may acidify a solution and then add hydrogen sulphide; in the next he may use zinc sulphate and add sodium hydroxide, first in small amount, and then in excess. The slight alteration in the appearance which the acid may produce will leave him in doubt as to whether it has any significance, and should be made a basis of inference or not; and if he decides that it should not, it may happen that the first effect of the sodium hydroxide will be so slight that he will neglect it also. Since the acid simply added hydrogen ions, while the sodium hydroxide produced a definite change, an interrogation-mark will call attention to the latter fact.

Sixth, some indication is necessary as to what is to be observed: for example, the use of the nose has to be enjoined many times before it becomes habitual in almost every experiment. Similarly it is sometimes necessary to draw attention to a change in colour which may represent a passing stage in a chemical change, or to the production of a gas which might be overlooked, as in the addition of a soluble carbonate to many salts of heavy metals. Perhaps separate exercises on these details of observation would be advisable, in order that the pupil may afterwards be left more completely to think for himself in later experiments.

Finally, definite questions should be asked in regard to the interpretation of what has been observed. These should be of two kinds which should be distinguished plainly from one another. Some it may be possible to answer from the observations and previous knowledge of the pupil alone; others may require reference to a book for part of the data. The pupil cannot tell, without some suitable indication, of which variety the question is, and will, in general, in every case make use of the book, and so miss the opportunity of thinking for himself, which the former variety of question would encourage him to do.[1]

It is evident, of course, that some mean must be struck between over-elaboration and too great compression of the instructions. They must not be so minute that to follow them will be wearisome, or so complicated as to be distracting and unworkable. If too concise, they will put more responsibility upon the teacher and pupil than the size of the class in the former case, or the intelligence in the latter case, will stand.

It is evident also that detailed directions will not be given in connection with every problem. In many cases the question will be stated and the pupil will be left to devise his own experimental method of attacking it and to do most of the thinking involved for himself. With large classes problems of this kind can be given only after much carefully directed work has been accomplished. With small classes, on the other hand, or when trained assistance is available, the more independent method may be used almost from the start and with the very best results.

---

[1] This question is discussed more fully under Use of the Text-book, chapter V., section d (p. 136).

c. *Selection of Experiments:* — In general the selection of the experiments should be made so as to afford the pupil opportunity to handle and become acquainted with all important substances, and to furnish him with material for systematic study of each topic in order that some material for generalization may be available. Such chemical changes only should be used as may surely be brought about when the conditions are definitely specified. The inclusion of a fact should be determined by its value for the purpose in view, and not because it is easy to show, or because its presentation is sanctioned by custom. The experiment should reach the point to be illustrated as directly as possible. Thus measurement of substances by observing the volumes of solutions is less direct than weighing, since the latter is the mode of measurement in terms of which chemical quantities are defined. Gravimetric experiments should therefore precede volumetric. Artificial methods should be replaced by natural when possible. For example, making sulphuric acid from sulphur dioxide obtained from sulphuric acid does not illustrate the commercial process, and is in itself stultifying. It is equally easy to burn pyrites (in a hard glass tube) in a stream of air drawn or driven over it.

Considerations affecting Selection.

Above all, the numerous limitations of the pupil, both in general and in view of his particular state of advancement, must not be forgotten. The apparatus which can be furnished by the laboratory or handled by the pupil must be thought of. The degree of skill and the knowledge which the pupil has acquired must be borne in mind. The length of the periods available for work must frequently lead to the exclusion of some valuable experiments. In discussing the results, the precautions taken by the pupil must be considered in the light of the much greater precautions which scientific work of permanent value demands. The small number of data he obtains, as compared with the mass of data which alone can furnish a basis for confident generalization, must be remembered. We shall presently discuss more fully the incompleteness of much of the pupil's work in consequence of these limitations, and the necessity for leading him to realize precisely how far he contributes to the result, and how far the book is to be called upon for furnishing an adequate foundation for the conclusion (p. 136, *cf.* also p. 99). All this will be made much clearer if some opportunity is taken to explain in detail some particular chemical investigation, with all the laborious purification of materials and analysis of multitudes of specimens which must be accomplished before even comparatively limited conclusions can be reached. Almost any account of inorganic research[1] will furnish material for this.

Intensive rather than Extensive Work.

The chief general rule is that the work should be, as far as possible, intensive rather than extensive. A sufficient sample of the whole ground covered by the science must be included, for there are many reasons which make this desirable in the course given in the secondary school. But it must be remembered that a thorough knowledge of one small portion really implies the ability to master other and different portions more rapidly, and is therefore, from every point of view, a thing desirable of attainment. The intensive method means also that the total acquisition must be greater, for it is only after we know something about some chemical substance that we are able to do the most intelligent work with it. Nor is intensive work more difficult than the other. On the contrary, it is much easier to enlarge our knowledge of a group of closely related things, than to enlarge it by passing rapidly from one group to another of things which are strange and less closely related. This instruction may tax the power of the teacher more, but it must be less difficult for the pupil. A Cook's excursion covering ten centuries of time and ten square miles of area in a single day is notoriously not the best means of studying the history and sociology of a people and its institutions. Laboratory work which resembles a personally conducted glance at many different things,

---

[1] For list of papers, see chapter VIII., section III. (p. 214).

may leave a confused sense of many more or less interesting impressions, but it cannot furnish an opportunity for learning chemistry.

It is this superficial quality which much school work possesses that prevents its recognition by the colleges. If the study of Latin in the school were of the same flimsy nature, and included no genuine investigation of the text, no mastery of the thing itself, and no adequate acquaintance with it in all its complexity as a medium of communicating thought, it would receive no recognition either. Both subjects must submit to the same test of educational value, or the whole work must be done over again in college. As Professor Bardwell says,[1] speaking of chemistry, "ingenuity and initiative power . . . come to the student not . . . by looking through experiments to greater things beyond, but by looking into experiments to find the simpler things which are near at hand." The same may be said of any of the benefits the study of the subject may confer. The limit of intensive study is to convert the whole work into research and give up the idea of covering much ground. My point is that both features must be preserved, and that of the two the former is the more important.

d. *An Illustration:* — In a well-known laboratory outline I find the following: "Treat a few small crystals of potassium iodide with concentrated sulphuric acid. What do you notice? Compare with the results obtained when potassium bromide and sodium chloride were treated in a similar way." This brief statement constitutes the whole directions for the experiment.[2] I have found this apparently simple experiment, at least at the stage at which it naturally appears in the course, by far the most difficult of the whole series for the year. In the first place, if

Potassium Iodide and Sulphuric Acid.

[1] New England Association of Chemistry Teachers, *Report of the Sixth Meeting*, 3.

[2] It ought to be said that the author of the book does not profess to offer the sort of directions combining instruction with direction which we have advocated. He says expressly that everything has been omitted that "does not serve to insure the success of the experimental work."

rather large crystals are taken, with much acid, and heat is not applied, no noticeable amount of gas may be given off at all. To get uniform results, the salt must be powdered and simply moistened with acid, and heating must be suggested in case the pupil does not get more results than he can take care of without this. In the second place, the pupil observes fuming in the air outside the mouth of the tube, a violet-coloured vapour in the tube, a brown film on the walls of the tube, an odour (sulphur dioxide or hydrogen sulphide, or both), and often a yellow sublimate (sulphur). Unless he is warned, he supposes that one body has all these properties. Without guidance he will never realize that from three to five distinct products are concerned. In the third place, he may not have yet met with free iodine, sulphur dioxide, or hydrogen sulphide, or, if he has studied them, he will have forgotten the properties of the last two. He certainly must have encountered hydrogen chloride, but he has probably forgotten that it fumed in moist air, and in any case he will not reason that, the halogens being similar elements, the new fuming gas must be hydrogen iodide. In the fourth place, he will try to put the whole of the products into one equation, and involve himself in an arithmetical puzzle of some difficulty, as well as a chemical absurdity. In the last place, he will fail to infer the oxidizing power of sulphuric acid and the easy oxidizability of the iodides, unless he is invited to do so.

I question the advisability of giving this experiment in elementary work at all. But in college work, to prevent the pupils being hopelessly muddled and discouraged, I have been led gradually to elaborate the directions. They have reached the following form, which will serve as an illustration of the sort of thing which is required to secure intelligent practical study of any problem: —

Model of Directions.

"Place about a gram of powdered iodide of potassium in a test-tube and moisten it with concentrated sulphuric acid (?). Warm, if necessary. Investigate the result as follows: —

"*a*. Breathe across the mouth of the test-tube to ascertain the effect of the gas on moist air. What gas previously made showed the same behaviour? Remembering the similarity between the halogens and between their corresponding compounds, what do you infer in this case? To confirm this conclusion, lower a glass rod dipped in ammonium hydroxide solution into the test-tube (?): also a strip of filter paper dipped in lead nitrate solution [R] (?).[1]

"*b*. What is the colour of the gas, or any part of it? What is the coloured body? (This assumes that iodine has been handled before.) Was there any corresponding product when sulphuric acid acted on a chloride? By what kind of chemical action could this coloured substance be formed from the one identified in *a*?

"*c*. Study the odour of the gas and describe it (?). Was there any effect on the lead nitrate which remained unexplained in *a*? Can you now explain it [R]? (This [R] assumes that hydrogen sulphide has not yet been studied.)

"The work in *a* and *b* leads to the recognition of two distinct gaseous products. That in *c* will yield one, and perhaps two others. Still another distinct solid product may be observed on the walls of the tube (?). Construct separate equations representing the formation of the first product from the original materials, and of each of the others from this product and sulphuric acid. What two properties of sulphuric acid and what property of hydrogen iodide are illustrated by this set of experiments."[2]

---

[1] [R] indicates that the pupil, in understanding what he is asked to do or in interpreting the result, needs information he cannot have gained in previous work, and must therefore refer to some book or to the instructor. Here he is ignorant of the action of iodides on lead salts.

[2] For examples of coherent directions and thorough working out of a problem, see the treatment of hydrates in Richardson, 6–8, of mechanical mixture and chemical combination in Remsen and Randall, 7–10, and of chalk in Perkin and Lean, chapters XIX. and XXX. A large proportion of the work in E. F. Smith and Kellar (*Experiments in General Chemistry*, Philadelphia, Blakiston) and Volhardt and Zim-

IV. The Pupil and his Attitude.

BIBLIOGRAPHY OF HEURISTIC TEACHING.

Second Report of the Committee of the British Association on The Present Methods of Teaching Chemistry. Report of the British Association, 1889. Also published separately by the Association. London. 1889.

Third Report of Same Committee, with additions by Professor H. E. Armstrong. Report of the British Association, 1890. Reprinted in NATURE, XLIII. (1891), 593.

Armstrong, H. E. On the Heuristic Method. Special Reports on Educational Subjects, vol. II. Printed for H. M. Stationery Office, London, and sold by Eyre & Spottiswoode. 1898.

The 46th Report of the Department of Science and Art. Printed for H. M. Stationery Office, London, and sold by Eyre & Spottiswoode. 1899. Abstract in NATURE, LX. (1899), 381.

Picton, Harold. The Great Shibboleth. SCHOOL WORLD, London, vol. I., Oct. and Nov., 1899.

Perkin, W. H. Vice-Presidential Address before the Chemical Section. Report of the British Association, 1900. Reprinted in NATURE, LXII. 477.

Discussion in THE SCHOOL WORLD, II. (1900), 358, 396, 437, 476. The last letter, by C. M. Stuart, gives an instructive illustration in detail.

Syllabus of an Elementary Course in Physics and Chemistry, issued by the Incorporated Association of Headmasters. London, Whittaker & Co. 1900.

Syllabus of an Advanced School Course in Physics and Chemistry, issued by the Incorporated Association of Headmasters. London, Whittaker & Co. 1899.

Whether we consider the best means of awakening and sustaining interest, or of fostering the scientific habit of thought, it is evident that leading the pupil to adopt the attitude of a discoverer will be most likely to accomplish the result desired. At the same time there are parts of the subject to which this method of approach is inapplicable. If, for example, we suggest that the pupil should discover the fundamental laws of the subject for himself, we are putting upon him an impossible task, and indeed deceiving him in regard to the nature of the foun-

---

mermann (*Experiments in General Chemistry*, Baltimore, The John Hopkins Press) illustrates these qualities admirably.

dation of a law. Verification is the term more applicable to work in this direction. Nor would we even suggest that the whole of the ordinary facts should be approached by the method of "find out for yourself," for the progress by this plan would be too slow for the purposes of a secondary school course. Much may be furnished in the class room, but the laboratory work should be divided between a small amount of verification and a large amount of what may be called investigation.

a. *The Verification of Laws:* — When the purpose of an experiment is the verification of a law, it will naturally be preceded by a careful study of the facts which the law covers.[1] While this necessarily carries with it full knowledge of the result of the experiment, it does not deprive the experiment of any of its value. Practical illustration will be required in order to make the understanding of the law more vivid, the recollection of its content more lasting, and, above all, to show by means of a sample, admittedly rough, what the general nature of its experimental basis is.

b. *The Attitude of Discoverer: the Heuristic Method:*[2] — It is evident that the nature of the directions will have much to do with the attitude of the pupil towards his work. In the ideal application of this method, however, no book and no directions are used. The questions to be solved are suggested as far as possible by the pupils themselves in the course of the examination of materials given to them. Naturally the demands made upon the pupil must be graduated, and at first the questions must be very simple.

Outline of Heuristic Work.

An outline prepared by Professor Armstrong (*Second B. A. Report*) will serve for illustration. Natural objects are examined, their origin, manufacture, and uses discussed, their appearance described. This is the first stage. Next, measurements of length, area, weight, density, temperature, and so forth

---

[1] Report of the Committee of Nine. University of the State of New York, *High School Bulletin No.* 7, 710.

[2] *Cf.* pp. 19, 54 and 56

are made. This is the second stage. Then the effect of heat on many things is examined for the purpose of gaining experience. Metals are heated in various ways; wood is dried and then burned for examination of the ash; minerals, such as sand, clay, sulphur, etc., are also heated. This is the third stage, and prepares for the fourth or problem stage in which the study of some chemical change may first be taken up.

The first chemical problem is that of determining what happens when iron rusts. The pupils must not only "find out for themselves," but as far as possible be led to imitate the detective's method, and find out *how* to find out for themselves. The question of the relative progress of rusting in moist and dry filings may suggest itself. Or air without water (dry) and water without air (boiled) may be tried. Then the iron may be weighed before and after rusting, and the search for the extra material begun. The question will be whether air or water furnishes the material. Moist iron tied up in muslin may be rusted in a pickle bottle inverted over water. The disappearance of part of the air leads to the treating of the same air with fresh filings, and of the same filings with fresh air, to see whether the change in either substance reaches a limit. The experiments may include other metals, and be extended in various ways according to the questions which suggest themselves to the pupils.

Nature, Limitations, and Results of Heuristic Work.

With a little care a series of interesting problems of this kind can be arranged in such a way that the solution of the preceding problems brings the pupil within measurable distance of the solution of the next. When sufficient skill has been acquired, the quantitative stage will be entered upon.

While in this way a large amount of chemistry may be learned, the object is not to teach chemistry, but to teach the pupils how to learn, — to confer ability and not knowledge. The progress in one sense will be slow, but the work, as long as interest is maintained and rational thinking and experimenting goes on, is fulfilling its mission. It cannot be objected that this kind of study is too difficult, for experience shows that it can be done even by young children. Picton (*The Great Shibboleth*) has outlined a course of this kind in chemistry for boys about twelve years old which he finds to work admirably. He says (SCHOOL WORLD, II., 397), "My experience is that the young boy of nine or ten can be readily got to think; the boy who has had considerable school training on ordinary lines can only rarely be got to think at all."

Illustrating the really remarkable way in which children, to whom text-books are quite unknown, will prove successful in the solution of intricate problems, a correspondent in the SCHOOL WORLD (II., 397) mentions four boys who, after two terms' work in a physical laboratory, investigated the rate of expansion of water when heated from 0° upwards. They used two methods, and got good curves for the apparent expansion. They saw clearly, however, the relation between this and the real expansion, as it would have been in the absence of the glass vessel. They tried first to measure the expansion of the glass with callipers between 0° and 60°, but, finding the results not exact, were at a loss how to proceed. When it was suggested that some Frenchman had determined the expansion of mercury independently of the vessel containing it, the boys ransacked the library and found a description of Regnault's apparatus and results. After being dissuaded from repeating his experiments, they corrected their own measurements by employment of his table. The writer adds, "undoubtedly the method has its drawbacks. The 'investigation' above mentioned occupied the better part of a term, during which, no doubt, the boys might have read through some little text-book, or pottered through a course of ready-made 'experiments' on 'heat.' It also cost the master . . . a good deal of labour. But he finds that a very little of this sort of work goes a very long way. . . . It seems to confer a power that is not acquired in any other way. The pupil's mind gains a freedom, a power of seeing things for itself, an alertness and adaptability in turning to fresh matter, which make great gaps in methodic knowledge of comparatively little importance. I have more than once been astonished at the ease with which boys, who have worked on this plan within a very small range, have been able to grasp the bearings of experimental work in quite another department, ... their eyes [seemed] to see things and processes in themselves, and not through the mists of conventional terminology."

Since the course of the pupil's inquiries must be as independent as possible, the direction which it may take cannot be foretold. The teacher must assist and guide with judgment, and, in general, as little as possible. A severe tax, however, will frequently be put upon the breadth of his knowledge of the subject, upon his time, and upon his mechanical skill.

Application in Teaching Chemistry.

While work exclusively on these lines, although pre-eminently suited to the needs of young pupils in the nature work of the grammar school and below it, does not furnish the knowledge of chemistry which is expected in the secondary school, it is evident that this attitude is the one to be cultivated when chemistry itself is being taught. I imagine that, when hydrogen is being studied, in nine hundred and ninety-nine cases in a thousand the pupil is informed directly or indirectly that it comes from the acid. Suppose that, instead of this, the question were raised whether the hydrogen came from the metal, the hydrogen chloride, or the water. If the pupil's experience had dealt with the last of these three substances only, the problem would be difficult to solve. If other metals were tried, many would be found to give hydrogen. Do these all contain it? Other acids would give hydrogen. Are they the source of the element? Substituting another solvent such as toluene prevents the appearance of hydrogen. Was water therefore the source of it? A little thought will show that a large amount of careful original work would be required to demonstrate that the acid was the source. In our teaching we are continually thus skimming airily over gaps which conceal not one or two steps, but whole flights of steps, all of which would have to be taken in a scientific study of the subject, although they are superfluous when memorization of the results is the only object.

Practically, the effort will be to include as much heuristic work as possible in the secondary school course. The spirit of it should certainly be of this kind. Now and then problems of a simple nature can even be given, after the material needed for their solution has been furnished, and the pupil may be left to provide his own directions for their solution. Thus, after equivalents have been measured, the pupil might determine the proportion of zinc oxide in zinc dust. Again, he might be instructed to find a solvent for some material; he might be told to make some salt in a pure form from an impure mineral, such as manganous chloride from manganese dioxide; or again, he might be asked to demonstrate the presence of one or more of the elements in ammonium carbonate. The exercises in the recognition of unknown substances (p. 178) are problems of the same order and are of the highest value.

The Ends of Instruction in Chemistry, and Means of Attaining them.

c. *Summary:* — It is safe to say that much chemical instruction does not reach the ideal sketched in this and the preceding sections of the chapter. Yet chemistry will never be recognised, nor will it deserve to be recognised, as Latin and other older studies are, either as an element in sound secondary education, or in work preparatory to college, until it is better organized along lines like these. We are only beginning to recognise this. It is not sufficient to suggest more thorough work. The pupil will be lost in details without the instruction which must go with it. Yet if the work is not made more elaborate, the result must be superficial. To fit the science for its place with the older studies, we must have guidance of a restrictive nature, which shall confine the possibilities of experiment, observation, and inference within limits somewhat like those set by the text, the grammar, and the dictionary. We must have guidance of an analytical kind to assist in the finding and study of all the points to be considered in each experiment. We must have guidance of a synthetic nature to stimulate the inter-relating of various facts and views brought out by present and past experiments. All this is necessary in order that the instruction may be an imparting of organized knowledge and not a jumble sale, and that, with the acquirement of an ever-tightening grip on the inner spirit of the science, rather than an ever-growing collection of rag-bag odds and ends, the pupil may advance in the profundity as well as the area of his knowledge. This alone can make chemistry a genuine means of culture and a discipline of real benefit in the later work of life. We need more detail, and at the same time more perspective. The Latin language cannot be studied by any other method; in this lies its strength. It seems to be possible to think that a study of chemistry which is not of this kind may still be a study of the science; in this lies its weakness. The purpose of scientific education is the application and higher cultivation of the critical powers by comparison, discrimination, and reasoning. It must also exercise and cultivate the power of scientific imagination, for, without this, no clear conception of the chemical tendencies of matter, and the conditions which influence their results, can be formed. That criticism and imagination are required in and are strengthened by its study, when this is prosecuted in the proper way, may be claimed for chemistry at least as confidently as for any other study.

V. Laboratory Technique.

One of the failings of chemistry teaching is the neglect of laboratory technique. The obvious value of neat and careful work, and of knowing how to adapt means to ends in mechanical matters, is so great, not only on account of its general educational value, but more especially because it is absolutely indispensable in really instructive chemical experimentation, that this neglect may well seem astonishing. It can be excused in any given case only on the ground that adequate supervision of a large class was impossible. In handling large classes of pupils who have already studied chemistry for a year in the secondary school, I have, for example, rarely found one who had any idea of how to ascertain whether a piece of apparatus was air tight or not. They usually blow into it as if it were a pair of bagpipes, oblivious of the fact that a hole nearly as large as their own throat would be necessary before the defect would be noticeable. The rational way of arranging the test, so that in some fashion the eye is the instrument used, forms an instructive lesson in itself.

Manipulation.

A good deal of attention is required in teaching proper manipulation. It is long before the pupil discovers that the stop-cock is meant for lowering the gas-flame, as well as for extinguishing it, yet he has continual opportunity to observe the risk in boiling a small amount of liquid in a large vessel with a large flame. It seems impossible to impress upon the minds of some pupils the proper method of folding a filter paper, of cutting it to circular form, and making it invariably smaller than the funnel. The clever use of the test-tube is a small art in itself. The pupil should learn the reason for the employment of different kinds of apparatus, such as retorts, flasks, test-tubes, etc., and in some exercises should be left free to select or devise apparatus for himself. The pupil is slow in learning the difference between thick and thin glass vessels in connection with the application of heat. Repeated misfortunes seem never to teach him that careful boring of corks and fitting of tubes takes no longer than making a funnel-shaped or ragged opening, and sometimes saves hours of time in subsequent work. The laboratory instructions, no matter how minute, will not secure the desired result without supervision and criticism by the teacher.[1]

Weighing.

Weighing, unless it has already been learned in the physical laboratory, requires careful preliminary instruction, if damage to the balance and discouragement in the work are to be avoided. The most frequent mistakes seem to arise from failure to count the weights correctly. Special emphasis should be laid on the necessity of ascer-

---

[1] These general operations are well described by Newth, *Elementary Inorganic Chemistry*, 15–34, by Young, *Elementary Principles of Chemistry*, Part II., 91–104, by Newell, *Experimental Chemistry*, 1–9, 329–353, and 365–369, by Peters, *Modern Chemistry* (Maynard, Merrill & Co.), 355–380, as well as by many other authors.

taining the weight, first by examination of the vacant places in the box, and then checking by counting the weights themselves as they are replaced. The working of glass, even if it go no further than the bending or drawing out of glass-tubing and fire-polishing of the sharp edges, requires a separate exercise. A Bunsen burner on which was inscribed in large letters, "do not use me in bending tubing," would be a boon to the teacher.[1]

Glass-working.

VI. Quantitative Experiments.

REFERENCES.

Newell, Lyman C. Quantitative Experiments in Chemistry for High Schools. SCHOOL SCIENCE (Monthly. Chicago, 2059 E. 72nd Place), I. 12. This new journal has already published several valuable articles on subjects of interest to teachers of chemistry.

Ramsay, Wm. Experimental Proofs of Chemical Theory for Beginners. London and New York, Macmillan. 1893.

Tilden, W. A. Hints on the Teaching of Elementary Chemistry. London and New York, Longmans, Green & Co. 1895.

Cornish, Vaughan. Practical Proofs of Chemical Laws. London and New York, Longmans, Green & Co. 1895.

Smith, Alexander. Laboratory Outline of General Chemistry. New York, The Century Co.; London, Geo. Bell and Sons. Fourth Ed., 1908.

We have already referred to the emphasis which is necessarily laid in chemistry upon quantitative measurement and the interpretation of the results. Imaginary examples, as we have hinted (p. 80), may serve when actual ones are not available, but the ease with which properly chosen measurements can be

[1] Clear instructions in regard to glass-working are given by Newth, *ibid.*, 35-39, by R. P. Williams, *Elements of Chemistry* (Ginn & Co., Boston, 1897), 384-387, and by G. M. Richardson, *Laboratory Manual and Principles of Chemistry* (Macmillan, 1894), 225-229. The teacher will find some accomplishment in this art invaluable. It is best acquired from direct instruction by some glass-blower. Much may be learned, however, by the study of works like Shenstone's *Methods of Glass Blowing* (Longmans, Green & Co., 1897), or Threlfall's *On Laboratory Arts* (Macmillan, 1898), chapter I.

carried out leaves little excuse for their omission, either from the demonstration or from the laboratory work of the pupil.[1]

a. *Limitations:*[2] — It is clear that the experiments chosen must be such that they are easily performed, and furnish fairly good results in the hands of beginners. They should employ no complicated or expensive apparatus. They should be capable of performance by a single pair of hands within the laboratory period. There is no disadvantage, however, in permitting two pupils to work together, provided they figure out the results separately. The most important condition is that it should be possible to furnish the pupil with instructions which will relieve the teacher of the burden of continuous supervision of each individual.

The chief misunderstanding which seems to arise in connection with this work is a confusion of it with quantitative analysis.

The Degree of Exactness Required.

The latter has for its object the learning of technique of the most refined description. The present experiments have for their use the comprehension of how quantities in chemistry are determined. Of course sufficient precautions must be taken to insure results which are approximately correct, or are at least concordant. It is frequently objected that results which are not exact are not only without value, but are misleading. This seems to rest on a misapprehension. No chemical work is absolutely exact. The conclusion always takes into consideration the sources of error, and the probable magnitude of the error, in applying the numerical value obtained. There is no reason why this should not be done in the experiments of beginners also. Indeed it should be one of the most instructive features of the work. Nor is there any reason why inexact results, within certain

---

[1] Their use is recommended by the Sub-Committee of the Committee of Ten, by the Committee of Nine, by the Committee on College Entrance Requirements, and, most recently, by the College Examination Board of the Middle States and Maryland.

[2] The whole subject of quantitative experiments is admirably treated by Dr. Newell in SCHOOL SCIENCE (see References).

limits, should fail to point to a law expressed in mathematically exact terms.

It is instructive to notice that most of the laws of chemistry were accepted long before they were confirmed by work showing any degree of exactness. Black (*Experiments upon Magnesia Alba*,[1] 1782), for example, converted 120 grains of chalk into quicklime and from this recovered 118 grains of the original material, showing an error of 1.6 per cent. Lavoisier decomposed mercuric oxide and ascertained the weight of the mercury and oxygen formed. The error appears to have been about one per cent, yet these results were held to furnish support to the law of conservation of mass. Proust ultimately triumphed in his controversy with Berthollet, although his own measurements of definite proportions showed errors varying from .5 to 5.5 per cent. The law of equivalent proportions was supported by Dalton by data which, in the light of modern work, are seen to be affected by inaccuracies sometimes amounting to 15 per cent. Dalton (*New Chemical Philosophy*, 318) quoted, in support of the law of multiple proportions, values for the ratios of nitrogen to oxygen in two oxides of nitrogen which show an error of 8 per cent.[2]

The ideal of quantitative work for beginners is 1 per cent accuracy. That this may easily be attained with suitable experiments, may be seen from the actual results of pupils' work in many schools where they are used.[3]

b. *Equipment for Quantitative Experiments :* — No elaborate equipment is needed for these experiments. Usually the same pieces of apparatus which are used in ordinary work will serve for them. The few special articles required may each be employed in several if not all of the experiments. A sufficient

---

[1] *Alembic Club Reprints, No.* 1. London, Simpkin, Marshall and Co.; Chicago, The University of Chicago Press. P. 29.

[2] This subject is discussed in detail by Vaughan Cornish. *Practical Proofs of Chemical Laws*, 15, 26, 43, 68, 79.

[3] Sample results are given by Newell in SCHOOL SCIENCE, I. 16, and on his *Teachers' Supplement*, 13, 14, 17, etc., and by Benton, SCHOOL SCIENCE, I. 148.

equipment for a large class does not imply that each member should be furnished with a complete outfit, since all need not do the same or any quantitative experiment at the same time.

The Balance. The chief item is the balance. Using an expensive instrument, however, is not only unnecessary, but wasteful. A balance with case, such as Becker No. 31, costing $15, and sensitive to one centigram, will serve all purposes. A set of weights (50 gr. — 1 cgm.), costing in a box $1.50, will also be needed. Newell (*Experimental Chemistry*, 347) describes a mode of enclosing common horn-pan scales, costing originally $1.25 to $2.25, which makes them applicable in this work, and other teachers confirm this statement.

One source of trouble lies in the rusting of the balance. This is reduced to a minimum in a form of the instrument which is manufactured entirely of aluminium and glass,[1] and is recommended and figured by Benton (SCHOOL SCIENCE, I. 148). Another source of annoyance is the continual loss of the smaller weights. This becomes impossible with the use of the Chaslyn balance,[2] figured on the back of the same number of SCHOOL SCIENCE (May 1st, 1901), in which rings which cannot be removed from the apparatus take the place of weights. I have found this balance very satisfactory.

Other Apparatus. The only other more or less special pieces of apparatus required are burettes (graduated, and holding 50 c.c.), porcelain crucibles (No. 0), porcelain boats, large bottles (one litre bottles, or five-pint mineral-water bottles), a barometer, and thermometers. Rubber stoppers save the loss of much time, and indeed are in the end cheaper than corks. Platinum ware is never needed, but clean crucible tongs will be found useful.

---

[1] Made by The Crowell Apparatus Co., Indianapolis.

[2] Made by The Chicago Laboratory Supply and Scale Co. Another form, "the triple beam balance," sensitive to 8 mgm., is made by the Apfel-Murdock Co. (82 Lake St., Chicago). A similar instrument is sold by Richards & Co. also.

c. *Suitable Quantitative Experiments:*—So many of these have been employed in recent text-books and laboratory manuals that detailed description is unnecessary. We may refer to a few which have been tried and found trustworthy. They are arranged according to the subjects in connection with which they are used.

Quantitative Experiments.

Definite Proportions:—Action of hydrochloric acid on varying quantities of ammonium hydroxide or sodium carbonate (A. Smith,[1] 12).

Combining Weights:—By direct union of copper and oxygen, or direct formation of cuprous sulphide (Tilden, 15). Indirectly by action of nitric acid on copper, zinc, iron, tin, or magnesium, and ignition leaving the oxide (Tilden, 14; A. Smith, 34). Indirectly by union of iodine and magnesium and formation of the oxide by ignition of the iodide (Young,[2] 34). By decomposition, mercuric oxide (Newth,[3] 105), silver oxide (Ramsay,[4] 97). The composition of water is somewhat difficult to measure on account of the small weight of the hydrogen (Newell,[5] 97; Tilden, 34; Perkin & Lean, 286).

Hydrogen Equivalents:—By measuring the volume of hydrogen displaced by zinc, magnesium, aluminium, sodium, etc. (Remsen,[6] 47; A. Smith, 35; Perkin & Lean, 204; Torrey,[7] 147). By measuring the weight of the hydrogen lost (Perkin & Lean, 206; Reynolds,[8] 23).

Inter-Equivalents of Metals:—Zinc and copper (Cornish, 91), zinc and silver (Newth, 140), magnesium and silver

---

[1] The names in parenthesis in this section refer to books listed in the Bibliography or described already in other connections.

[2] A. V. E. Young. *Elementary Principles of Chemistry.* Part II.

[3] Newth. *Elementary Inorganic Chemistry.*

[4] Ramsay. *Experimental Proofs of Chemical Theory.*

[5] Lyman C. Newell. *Experimental Chemistry.*

[6] Remsen & Randall. *Chemical Experiments.* New York, Henry Holt & Co. 1895.

[7] James Torrey. *Studies in Chemistry.*

[8] J. E. Reynolds. *Experimental Chemistry.* Part I. London and New York, Longmans, Green & Co. 1897 (7th ed.).

(Reynolds, 17), iron and copper, magnesium and silver (Perkin & Lean, 302 and 305).

Multiple Proportions:—Reduction of cupric and cuprous oxides (A. Smith, 38). Reduction of nitrous and nitric oxides and collection of the nitrogen (Ramsay, 82–86). The reduction of lead monoxide and dioxide will be found suitable if the pure substances can be obtained. Note that the former takes up carbon dioxide from the air. The monoxide is difficult to reduce.[1]

Solubility of Salts:—Measurement at different temperatures (Richardson, 9).

Raoult's Laws:—Depression of freezing point and elevation of boiling point of solutions (Young, Part II., 54–57).

Gas Density:—Several excellent methods are described by Professor Ramsay (*Ibid.*, 26, 34, 39, 45). These have been borrowed freely, and many of them will be found in the other books we have quoted. Another method, that of Regnault (Perkin & Lean, 234), gives good results.

Volumetric:—This takes the form usually of titration of solutions of acids and bases. Volumetric experiments with gases, illustrating Gay Lussac's law, we owe chiefly to Hofmann. These are concerned with the volumetric composition of steam, ammonia, and hydrogen chloride; they are described in many works. The combination of oxygen and nitric oxide (Tilden, 252; Young, 40), the volumetric composition of nitric oxide (Tilden, 251) of ammonia (Ramsay, 59), and of the air (Cooley,[2] 61) will be found useful.

Special:—A very instructive experiment, in which a weighed amount of silver foil is converted first into the nitrate, then into the oxide, and finally back to silver, is used by Benton (SCHOOL SCIENCE, I., 157). It has the advantage of enabling the pupil to check his result, since the silver is weighed at the

---

[1] Other instructive illustrations are given by Young (*Ibid.*, Part II., 30), W. R. Smith (SCHOOL SCIENCE, I., 87), A. Smith. 18.

[2] Le Roy C. Cooley. *Laboratory Studies in Elementary Chemistry.* New York, American Book Co. 1894.

beginning and end. The reduction of silver nitrate by hydrogen (Cornish, 34), measurement of water in hydrates ('water of crystallization,' A. Smith, 24), and the proportion of the carbon dioxide in a carbonate (Newth, 228; Newell, 215) will also be found applicable. The determination of the composition of zinc chloride (Torrey, 140), when taken in connection with the measurement of the hydrogen equivalent of zinc, permits a complete investigation of the action of zinc on hydrochloric acid to be made.

d. *The Application of Quantitative Experiments:*—There is one danger to which the use of exact measurement is liable, and that is, that the pupil may be misled into thinking that the operation of measurement is an end in itself. The scientific mechanic who cannot see beyond the cross wires of a telescope is not the person we are trying to train. Measurement is a tool and should be used, aside from a preliminary exercise or so, only in the solution of some definite problem. As Professor Perkin[1] says, "measurements should, in fact, be made only in reference to some actual problem which appears to be really worth solving, not in the accumulation of aimless details." It is in this respect that these experiments resemble investigation rather than quantitative analysis.

Time of Introduction.

The time at which the first quantitative experiment may be given naturally depends upon many things, particularly the previous experience of the pupil. Some practice in ordinary chemical work will be needed by way of preparation. The experiments should be used, however, not later than the laws which they illustrate, and measurements of combining weights must certainly be introduced before equations are used. To leave them to the end of the course is practically to postpone them until they become superfluous. Their early introduction is particularly desirable, in order that the pupil, in spite of the laws he may have learned, may not acquire from

[1] Vice-Presidential address already mentioned. See on this point Picton, *The Great Shibboleth* (SCHOOL WORLD, October and November, 1899), and also Lean (*Ibid.*, II. (1900), 78).

his practical experience the impression that chemical proportions are after all purely matters of chance. The teacher can only find out by trial the earliest point at which, with his particular class, they may be introduced.

Treatment of Poor Results.

When obviously inexact results are presented, they should never be dismissed abruptly and with contempt. Sometimes a discussion of these very results, and how he got them, with the pupil, will teach more than if they had turned out well, and had been accepted without criticism. The fact must be continually impressed on the mind of the pupil that it is the conscientious performance of the experiment that is wanted, and not a certain result. If the reverse impression is given, the pupil may resort to 'cooking' his figures, and the exercise may do harm instead of good. The teacher should always ascertain for himself, by trial with the same apparatus, the limits within which results may be accepted as representing good work.

In all cases the pupil should be warned not to throw away the product, in case the result seems to be bad, until he has submitted it to the teacher. Sometimes the result may be corrected, and repetition of the experiment be avoided, as when through misunderstanding the pupil gets a result, correct, but different from that which he had expected; when the product has been insufficiently dried; or when some arithmetical error has been made in the calculation. Pupils rarely feel any reluctance to repeat experiments of this kind, a fact which in itself testifies strongly to the interest they feel in them.

e. *Benefits and Objections:* — The general benefits which these experiments confer scarcely need enumeration. They

Benefits.

teach the necessity for care, exactness, patience, and cleanliness, by themselves demonstrating too often the effects of lack of application of these elementary virtues. They give the pupil a confidence in the exactness of the experimental basis on which the science rests, and a respect for exact experimental work, which he could not otherwise attain. They take time, but their very slowness is in some ways an advantage. The laboratory should be a place for thinking as well as for seeing. I have found that questions often suggest themselves to the minds of the pupil during the leisure which some stages of these experiments permit, the effort to answer which teaches them much they might not have otherwise learned. The arithmetical problems arising out of these experiments, founded as they are on their own data, are worked by the pupils with an amount of interest, not to say eagerness, which artificially made problems can never inspire.

Objections.

Some of the objections[1] which have been urged against their use have already been noticed incidentally. The statement that high school pupils lack skill to carry out these experiments is either a commentary on the selection which the teacher has made, or a piece of rather obscure humour. It is in the effort to gain skill which they call forth that part of their value lies. The argument that in colleges quantitative analysis usually does not appear until the third year, and that quantitative experiments are not given in general chemistry, may be a criticism of college teaching, but it is not an argument against the use of these experiments. Finally, the suggestion that historically chemistry was qualitative before it was quantitative, and that the historical order should be followed, seems to misapply an important principle. The history of modern chemistry begins with Priestley, Lavoisier, and Cavendish, but it was the quantitative part of their work which alone really deserved the designation fundamental. It is difficult to see why a pupil should be dragged through a fog-bank of alchemy and empiricism simply because the rest of the world lost its way and wandered in such a fog for hundreds of years.

---

[1] An extended treatment of a long list of objections, including all that have been urged with the exception of two, is given by R. P. Williams, New England Society of Chemistry Teachers, *Report of the Fifth Meeting*, 3–6. Some of the arguments in their favour are well put by Young, *Suggestions to Teachers*, designed to accompany his *Elementary Principles of Chemistry*, 2–4.

VII. The Rôle of the Teacher in the Laboratory.

Continuous Attendance in the Laboratory.

One of the most serious faults of much chemistry teaching is that the pupils are allowed to work by themselves in the laboratory in the absence of the teacher. None of the benefits we have enumerated above, or of the results anticipated from the methods of laboratory instruction just described, can possibly be realized in the smallest degree when this course is pursued. The pupils cannot be expected to teach themselves chemistry any more than they could give themselves instruction of the slightest value in Latin or mathematics under the same circumstances. The Latin room cannot teach Latin, and the chemical laboratory is not more fit than any other apartment to take the place of the instructor. The natural result of neglect of continuous and strenuous supervision is that the pupils think that the performance of prescribed mechanical operations constitutes a study of chemistry. This tendency of all laboratory work is exceedingly difficult to combat, and continual questioning by the teacher can alone keep the work on the level of an intellectual exercise. No laboratory outline, however carefully prepared, can take the place of the living teacher. His questions are directed to the particular features of the particular way of doing each experiment and to the particular misconceptions or shortcomings of each pupil. No two cases are ever precisely alike, and therefore no printed questions can ever meet the difficulty. The disastrous blunder of permitting or encouraging unsupervised work seems to be commoner in colleges than in secondary schools. But, until it is recognised and remedied, we can never secure either culture or a knowledge of chemistry, either for the ordinary student or the prospective specialist, merely by including of the science in our curricula.

Chemical manipulation is an art. It cannot be acquired without models to copy and trenchant criticism as the work proceeds. The latter must be applied the moment occasion for it arises, or hours may be wasted in trifling with unimportant features of an experiment, or in using an imperfect or inadequate piece of apparatus. Supervision of the technique is as necessary in chemistry as in drawing or shopwork. Some pupils seem naturally to possess the 'knack' of working neatly and successfully with little assistance, but these are very few in number. The great majority are utterly incapable of giving concrete expression to the directions without frequent suggestions and warnings.

Teaching Technique.

At the beginning, one teacher cannot handle successfully more than fifteen students. The more the number assigned to him exceeds this, the less thorough the instruction and the longer the time taken in reaching the same degree of proficiency must be. When once a good start has been made, equally efficient work may be done with a larger proportion of pupils to each instructor. If a sufficient force of instructors is not available, the work can be simplified and more time can be taken in covering the same ground.

Ratio of Instructors to Pupils.

VIII. The Note-book.

REFERENCES.

Arey, A. L. A Paper on the Management of Laboratory Classes in Chemistry, and the discussion following its reading. Albany, N. Y., The University of the State of New York. High School Bulletin No. 7 (1900), 678–684. This covers almost all phases of the subject.

Cooke, J. P. Laboratory Practice. Pp. 6–8.

Keeping a note-book is a valuable aid in laboratory study. The notes should be provided with prominent headings indicating the part of the subject which is being studied and the object of each experiment. Following this should appear a statement of what was done, including the materials used, a description of the apparatus (with a sketch, if it seems called for), and the procedure adopted. When all this is detailed in the laboratory directions, however, it does not seem necessary that it should be repeated, unless perhaps in an abbreviated form. Next, the observations which have been made

Content.

should be stated, then the inferences drawn from these, and in most cases the chemical equations representing the changes should be given.

Form.

Care should be taken in regard to the form in which the notes are presented, but the lavishing of too much time upon the unnecessary copying and beautifying should be discouraged. The use of concise yet clear English should be imperatively demanded. But a too formal division of the notes into columns containing "requirements, conditions, observations, conclusions," is not sufficiently elastic, represses the individuality of the student, cultivates a mechanical view of the subject, and should be avoided.

The majority of teachers favour the writing up of the notes in final form in the laboratory rather than at home. This is undoubtedly the better method. Inasmuch as attainment of the best form cannot be reached in this way at once, it is well to use the even folios of the book for memoranda and ciphering, and to write the notes in more formal fashion on the odd folios opposite, and to do this immediately after the experiment has been performed.

Examination of Note-books by Teachers.

It is indispensable to the success of the system that the note-book should be examined periodically by the teacher, and all blunders in English, errors in observation and mistakes in chemistry marked distinctly. The corrections themselves, however, should by no means be made by the teacher. In discovering the truth and making the necessary change himself, the attention of the pupil is called to the matter much more forcibly. The note-book should be examined immediately after the first exercises, in order that by criticism and suggestion the best way of making the notes may be most quickly communicated. Later they should be examined at regular intervals. Some teachers require that the note-books be left in the laboratory at all times, and provide a shelf near the door on which they may be filed as the pupils pass out. They are thus available for examination during any moments of leisure which the teacher may find.

The reading of note-books when the class is large is the most laborious and least attractive task of the teacher. Indeed, in many cases, systematic examination of all the books by one person is impossible without assistance. Sometimes a classroom hour may be devoted to the reading of notes by some of the pupils and criticism by the other members of the class. Often former pupils may be induced to take a share in the work. In Normal Schools, in fact, the students will receive distinct benefit from an opportunity to assist, to some small extent, in the instruction by examining note-books and taking part in the supervision of the laboratory work.

Value of the Note-book.

The extreme value of keeping a note-book in a suitable style cannot be doubted. It impresses the facts ten times more strongly on the memory than would be the case without its use. It gives practice in accurate and clear expression. As an incident to the writing, the pupil usually finds his thoughts on the subject were not so perfectly organized as he had supposed. In framing written answers to the interrogation points and questions in the directions, he is stimulated to group the facts in new ways, and is assisted in studying the subject by the discovery of gaps in his thought and in his observation which otherwise would have passed unnoticed. If the note-making is to be perfunctory, it had better not be attempted at all, for, instead of yielding the benefits we have mentioned, it will simply waste the time of both pupil and teacher.

IX. Emergencies.

Danger from Injuries.

Guarding the pupils from injury by specific laboratory directions,[1] due and pointed warning, and continuous oversight is one of the most serious responsibilities of the teacher of chemistry. When, in spite of this, slight accidents occur, as they frequently do, he must be prepared to treat the

[1] As prevention is better than cure, the pupils should be positively forbidden to make any experiments of their own devising without first consulting the teacher.

injury properly. Aside from damage to the eyes, burns are the most serious injuries with which he is called upon to deal. They are to be regarded very seriously, because, through the destruction of the protective power of the skin, infection will almost always occur unless the burn is very small indeed. This will be followed by suppuration, and the resulting wound will leave an exceedingly ugly scar. In such cases, therefore, careful disinfection should never be omitted.

Burns.

Burns through contact with hot bodies, or from burning liquids like alcohol, should be treated first with an emulsion of linseed oil and lime water. Burns produced by corrosive liquids like bromine, sulphuric acid, and nitric acid should be washed with water, and then the part should be rubbed gently with a paste made of sodium bicarbonate and a little water (the normal carbonate is alkaline and, having an irritating effect, should not be used). In all these cases, to prevent infection, carbolated vaseline or powdered boracic acid should be applied liberally to every part of the surface burned, and a bandage should then be wound around the whole. A "wet dressing" is often used. Saturated boracic acid solution, (1 : 20) diluted with an equal volume of water, is employed. The piece of lint, large enough to extend some distance beyond the burn in every direction, is soaked with this solution, and covered with a sheet of oiled silk or "protective" to restrain evaporation. Burns caused by phosphorus are the most difficult to heal. They should be first cleansed by washing with a brush dipped in water containing a little carbolic acid. If necessary carbon disulphide may be applied. The best results seem to be obtained when the wound is then powdered over with picric acid and wrapped in a wet bandage. When the injury includes the contact of acid with the eyes, washing with water and a solution of sodium bicarbonate projected from a wash bottle should be applied, and the victim of the accident sent at once to a competent physician.

Cuts should be washed out with water, and, after certainty has been reached that any glass they may contain has been removed, they should be covered with court-plaster. A solution of 'iron persulphate,' or, in an emergency, ferric chloride will arrest bleeding. If the cut is otherwise than small, a disinfectant will be required. A dry mixture of salicylic acid, one part, and boracic acid, two parts, applied liberally and held in place by a bandage, is a suitable dressing. In case of faintness, inhalation of ammonium hydroxide, or administration of five drops of ammonium hydroxide in a little water will usually be effective. The irritation caused by inhaling acid fumes will be relieved by inhalation of ammonia, and that from chlorine and bromine by the inhalation of vapour of alcohol.

Cuts.

Fires caused by burning liquids like carbon disulphide are not affected by water, and should be put out by liberal use of sand. Burning clothing can be extinguished best by means of a wet towel.

Fires.

The various materials mentioned above, along with a pair of scissors, should be kept on hand in some special cupboard, in a conveniently accessible position, and they should never be used for any other purpose than that for which they are intended.

CHAPTER V

INSTRUCTION IN THE CLASSROOM

In order that the purposes which we have so far explicitly discussed, or implicitly assumed, may be realized, several distinct means of instruction are at the disposal of the teacher and should all be used. Of these the individual laboratory experience of the pupils is the most important. The utilization of this experience, however, will never occur spontaneously. The results will remain largely incoherent and meaningless without discussions, — 'quizzes' — in which they are infused with life, experimental demonstrations in which they are amplified, problem-working in which they are made more definite and are driven home, and book study and reference work in which they are brought into relation with the rest of the science.

a. *Oral and Written Quizzes:* — The oral quiz naturally follows the laboratory work and deals mainly with this, because it is the noting of the significant facts and their translation into chemical knowledge which gives most difficulty to the beginner. It will draw out much that was unheeded at the time, but remains accessible to careful questioning, and so will prepare the way for more adequate observation in the future. It will also relate this work to the statements of the book and keep the two from remaining two different things, as they have a tendency to do. Through criticism of loose expressions, by the teacher and by other members of the class, it will bring out lack of clearness of thought and at the same time teach discrimination in the use of language.

The Services Rendered by the Quiz.

Aside from these, there are assigned to it three services which would remain entirely unrendered if the quiz did not undertake them. One is that of developing the generalizations of the science from the facts, of which those observed in the laboratory are samples. Thus the pupil may have treated zinc with haif-a-dozen acids, yet will almost never even speculate on the probable generalization unaided. The second service is in practising the application of the generalizations to chemical questions and, when possible, to those of every day-life. Generalizations are the tools of thought, and unless they are put to some use the labour involved in their manufacture will have been largely wasted. The third service is in exercise of the scientific imagination, without which attainment of even the slightest degree of chemical intelligence is impossible. This furnishes one form of the so-called 'explanations' (*cf.* p. 147) which, when legitimately used, are so helpful.

To sum up, the object of the quiz is to lead the pupil to gain the scientific habit of mind by practice in the scientific treatment of a specific science.

Of these features of the quiz, only the two last seem to demand special discussion.

Service in Showing Application of Conclusions.

When a generalization has been stated it will find immediate application. Frequently some little time will have to be devoted to making the application plain. For example, the law of conservation of matter finds illustration in the results of raising the same crop on the same piece of land year after year. If the product is one which is cut and carried off entirely, the constituents of the soil which are essential parts of the food of the plant are effectually removed. Analysis of the soil and of the plant show at once what stock of plant food is available, and how long it will last. The use of fertilizers and other expedients replaces or brings within reach of the plant the phosphates, for example, which are indispensable to its growth. If it is the law of definite proportions which is under discussion, illustrations are abundant. In its absence we could not regulate the heating of our houses, because with the same draft and supply of oxygen the combustion would be more fierce at some times than at others: we could not make a contract for the supply of iron because we could not foretell what amount of coal would be required to reduce our ore, and therefore what the expense of producing the metal was likely to be: we could not offer photographs at so much a dozen, because the second half of the dozen might cost a thousand times as much to print, develop, or tone as the first. Commercial analysis, by the results of which values were to be determined, would be made utterly in vain. In fact, the conduct of all industries depending on chemistry would be impossible as business enterprises. Even life itself would cease, since its continuance depends on the assumption that approximately constant quantities of food will give approximately constant results in the way of nourishment. A little thought will show that similar illustrations of almost all the generalizations of chemistry may be found. Visits to factories will usually furnish many opportunities for pointing out applications of facts noted in the classroom.[1]

Service in Furnishing Explanation.

By inference is meant a rigidly logical process in which no steps are omitted and no gratuitous or surreptitious additions are made to the conclusion to which the data fairly lead. For example, when hydrogen chloride is formed by the action of sulphuric acid on salt, we infer that under the circumstances the hydrogen of the acid could unite with the chlorine and the sodium with the sulphanion (SO_4); that is, that affinity between these materials exists. We may *not* infer that this affinity was much greater than that which held the original compounds together, nor that sulphuric acid is more active ('stronger') than hydrochloric acid.[2] Nor may we infer that a tendency to the formation of gases accounts for the action. 'Accounting for' and 'explaining' chemical changes is a risky proceeding. It is usually beyond the beginner. In this illustration, a knowledge of mass action is needed for the purpose. Supposing causes should never be

[1] This subject is discussed farther in par. e, p. 138.

[2] As a matter of fact, both these conclusions would be completely erroneous.

indulged in. Explanations are very satisfying, but we must be careful to avoid incorrect ones as they are much worse than none. An illustration of the use of the imagination will help to show how far the attempt may safely go.

Service in Employment of the Imagination.

The imagination (*cf.* p. 11) must be applied to everything in chemistry. For example, why is the generation of chlorine such a leisurely process? It is not for lack of affinity that the interaction of the manganese dioxide and hydrochloric acid refuses to be hurried even by a blast-lamp! To answer the question, we have to imagine the whole affair in detail. The molecules of the substances must meet to act. The acid is in solution, and from six to twelve molecules of water encounter a lump of dioxide for every one of acid that reaches the goal. After the one acts, another has to come up by diffusion, a slow process. Again, the dioxide is insoluble and does not go to meet the acid. How great is the contrast between the action of hydrochloric acid on similar pieces of marble and of sodium carbonate on this account. The molecules of dioxide have to be sought, and only the surface ones, the merest infinitesimally small fraction of the whole, are within reach. Contrast this with the rapid action in the 'Seidlitz' powder,' where both bodies are dissolved. Then the manganous chloride formed has to diffuse away to expose a new surface. And, when we try heating, we cannot raise the temperature much, because aqueous hydrochloric acid boils at 110° or lower. Contrast this with our custom of raising the temperature to a red heat in making oxygen. Here, to avoid distilling over some of the acid, and so wasting it and rendering the chlorine impure at the same time, we may not go even as high as 100°. It will be noted that we are not accounting for the chemical action, or supposing causes for it. We are simply considering the details and trying to explain how the conditions affect the change. The tremendous rôle which the imagination plays in this hardly needs to be pointed out.

"Imagination is thought by means of images" (Wundt). It gives new form or grouping to the relations of the contents of the memory and the percepts of the senses. In the above illustration it uses the pictorial imagery of the molecular theory, a multitude of facts, and some ideas about molecular forces for the production of a rationalized kinetoscope picture of the whole proceeding.

The quiz will fitly occupy a large portion of the whole time near the beginning, when all is new, and again during the last half of the course. As the subject advances, earlier matters, already partly forgotten, receive fresh light from and reflect valuable light upon each successive topic.

Some of the objects of the quiz enumerated above will be especially well served by occasional written exercises or informal examinations. These are particularly valuable inasmuch as they give the pupils practice in making connected statements, such as accounts of the properties of substances and extended discussions of chemical questions, and so train them in the chemist's way of classifying his facts and expressing his conclusions. They also furnish occasion for that continual reviewing which is so indispensable.

Written Exercises.

The nature of the questions asked in a written exercise shows more clearly than any other one thing the kind of instruction they are testing. Questions such as: What are the colours of the precipitates when such and such substances are mixed? or, Give the graphic, semi-graphic, and empirical formulæ of the following substances, — are tests of memory and show serious misdirection of the pupils' energy. The questions should test the powers of reasoning, discrimination, and co-ordination, as well as the knowledge of the pupil. They should be constructed so as to demand reference to laboratory experience for correct answer.

For example, if we ask what the action of hydrochloric acid on quick-lime is, the answer may be given by rote. If we ask, How would you show the presence of oxygen in quick-lime? the pupil's thought, disciplined by laboratory experience, alone can furnish the reply. The answer, "I don't know," or "I don't remember," which we often receive, is a pointed commentary on the habit of mind our educational methods seem to engender. The question invited the pupil to think, and this was so unusual that he did not even recognise the fact.

The Committee of Ten recommended that examinations should be practical as well as oral or written. They referred to college admission examinations, but the idea is equally applicable to any test of acquirement whatever its purpose.[1]

Practical Examinations.

b. *Experimental Demonstrations:* — Some teachers prefer this work to precede, others to succeed the laboratory exercise on the same topic. In the latter case the desire is to let the pupil examine the subject first entirely by his own efforts. This is, doubtless, as a general rule, the best plan. But, while the pupil is still ignorant of the handling of apparatus and the kind of phenomena to be expected, it must involve slow progress and much supervision. After some experience has been gained, it is undoubtedly more instructive as well as more interesting than the other arrangement.

The first of the uses of the demonstration is, in connection with the earlier exercises, to show simple experiments, to point out the matters of observation and to indicate the inferences to be drawn. These in fact will be model laboratory studies, showing something of the nature and use of the apparatus, and intended to save the pupil much needless bungling in his first efforts. The experiments need not be the same as those performed in the laboratory. Yet even if they are, the whole affair seems so

Uses of the Demonstration.

---

[1] Perkin and Lean give a large number of simple problems (*ibid.*, 324–326) to be solved by practical work in the laboratory, which, even if they are not used for examination purposes, will nevertheless afford hints that may be utilized in other ways. The recent examination papers of the University of the State of New York will assist in showing what are deemed the most important things in the science. Ellis' *Papers in Inorganic Chemistry* (London, Rivingtons; New York, Longmans), containing eight hundred questions and problems, with numerical answers to the latter, and a volume of *Questions on Chemistry*, by Jones (Macmillan), may be found useful. Sets of questions in chemistry are published every month in the SCHOOL WORLD (London).

different in one's own hands from what it appears with an expert at the helm that more than enough remains to be learned to repay the repetition. There will be so many physical considerations of a strange kind connected with the apparatus and the chemical substances, that these alone, quite apart from the chemical facts, make the first laboratory exercises sufficiently hard in spite of the utmost assistance the teacher can give.

Experiments requiring special skill will usually be shown by the teacher. To put these in the hands of beginners would be to invite failure and discouragement. Of this nature are the experiments of Hofmann[1] on the law of volumes.

Never show Untried Experiments.

Experience is nowhere more needed than in this work.[2] Yet, whatever his experience, the teacher should never show an experiment he has not tried with precisely the *same apparatus* and *materials* he intends to employ. Different lots of the same substance are not always identical, and even a lot previously used will deteriorate and cease to be trustworthy. Care in these matters is usually learned only after several humiliating experiences. The teacher will also find it difficult at first to see the experiment as his pupils view it, to put himself back in their place and omit nothing essential in making clear the construction of the apparatus and its working, to draw attention to every feature in the progress of the whole operation, and finally to wring from it the lessons it teaches to the last drop. In all this, the interest of the

---

[1] See Smith, *Laboratory Outline of General Chemistry*, pp. iii–iv. It is a great pity that the English edition of Hofmann's delightful *Introduction to Modern Chemistry* (London, 1865) is out of print and difficult to procure. It contains the best models of experimental lectures, both as regards presentation and illustration, extant. The German translation (*Einleitung in die moderne Chemie*, Braunschweig, Vieweg, 1877) has reached its 6th edition.

[2] See Newth's *Chemical Lecture Experiments* (Longmans, Green & Co., London and New York, 1899). Also Benedict's *Chemical Lecture Experiments* (Macmillan, London and New York, 1901). These works will be found indispensable, even to the practised experimenter.

With few exceptions, the apparatus used in demonstrations should be the same as that used by the pupils in similar laboratory experiments.

class, which never flags when anything is going on, may be utilized and directed so that, in response to questions put by the instructor and by themselves, most of the points just mentioned are covered.

Demand on the Time of the Teacher.

The preparation of experiments consumes much time and requires some ingenuity. Keeping the demonstrations up to the highest standard that the equipment of the school permits demands heroic effort, and often some self-sacrifice on the part of a busy teacher. School authorities have for the most part still to learn that a teacher of science cannot carry as many hours of classroom appointments with efficiency as teachers of most other subjects. As Dr. Newell says,[1] he "needs time to arrange the workshop of his class; time to consult with individual pupils; time to repair, clean, arrange, and replace apparatus; time to clean up what pupils and janitors will not do; time to mix solutions, put them in properly labelled bottles, and the bottles in the customary place; time to correct laboratory notes and see that the pupils understand the corrections; time to arrange lecture experiments and remove the unsightly results before the room is again used; time to visit with classes the neighbouring shops and manufactories which illustrate the industrial phases of chemistry; time to read current scientific literature; time to *rest* physically and mentally, so that he may come daily to his classes with that mental poise which is essential to successful teaching." Reasonable time within school hours for most of these tasks is as necessary for good teaching of chemistry as the materials and laboratory themselves.

Problems.

c. *Stoichiometric Problems:*—The working of problems in considerable numbers by individual pupils seems to be an exercise too often neglected. This is acknowledged to be a valuable aid in enforcing the quantitative character of every chemical change, and in holding the pupil's attention on, and making him familiar with combining weights and their use. More difficult problems, concerned with the

---

[1] SCHOOL REVIEW, IX. (1901), 288.

calculation of molecular and atomic weights, and other allied subjects using the laws of gases, form the readiest means of clinching what may otherwise remain a mass of loose and ephemeral ideas. Sample cases should be worked in the classroom. After the pupil's exercises have been corrected, it will be found advisable to discuss them with the class.[1]

d. *Use of the Text-Book:* — Some teachers prefer to use no regular book, and instead refer their pupils to certain passages in works contained in the school library. Their pupils, however, generally lack fulness in knowledge of the subject. To throw the pupil on his own resources is an excellent idea, but this plan seems to carry it too far. Reading in other books, however, is also highly advisable, if there is opportunity for it. In any case, turning of the work into humdrum preparation of so many pages of printed matter daily is easily avoided.

It is well to have some familiar source to which the pupil may turn for assistance in recalling old matters. There is some advantage also in the pupil's becoming perfectly acquainted with the arrangement of one book, which shall employ approximately the order followed in the laboratory. This helps him in getting a more definite grasp of the relations of the parts of the science.

The chief reason for the use of books, and preferably in the main, one book, lies in the fact that there will hardly be a single chemical change of which the pupil can make a complete study, if he is thrown absolutely on his own resources. His own work furnishes him with a part only

A Text-Book necessary.

[1] A graduated series of problems, with answers, is given in Whiteley's *Chemical Calculations* (London and New York, Longmans, Green & Co.). Waddell's *Arithmetic of Chemistry* and Lupton's *Elementary Chemical Arithmetic* (London and New York, Macmillan) are similar books. Many problems will be found also in Newell's and in Perkin and Lean's books already mentioned, in E. F. Smith and Kellar's *Experiments in General Chemistry* (Philadelphia, Blakiston) and in Tilden's *Introduction to the Study of Chemical Philosophy* (London and New York, Longmans, Green & Co.). The teacher will find an admirable series of problems in physical chemistry in Bräuer's *Aufgaben aus der Chemie und der physikalischen Chemie* (Leipzig, Teubner, 1900).

of the information necessary for reaching the chemical conclusion to which his experiment points. If, for example, he burns phosphorus in oxygen, he sees a white cloud. He may succeed in showing that this is a solid, that the gas is used up, and that the solid is the only product. These things are within his powers of observation, although the experiment seldom seems to be carried even as far as this. But even so, he can only infer that a solid compound of the two elements has been formed. His study of the problem has been suspended before it could be clinched, on account of the difficulty in measuring the composition of the product. Even if he could do this, however, he would still require to get the combining weights from the book. Usually he gets both, in the formula, from this or the teacher. I have seen many note-books in which the observation of white smoke and the inference that the "product was P_2O_5" were the sole entries. It is worse than waste of time to encourage the pupil to make sham inductions like this from ridiculously inadequate data. When he has been told explicitly that his laboratory work is not expected to furnish all the necessary information, he appreciates the possibility of extending this information by more elaborate experiments. If, on the contrary, he is left in doubt on this point, and yet is asked to write the equation for every chemical change (*i. e.* to draw a quantitative conclusion from qualitative data) he will perceive the existence of an imposture, even if he cannot point out where it lies.

This is not criticism of an unreal state of affairs. Much chemistry teaching is like setting out to invite a man to test a stair, and then almost carrying him up bodily. He not only learns nothing about the soundness of its construction, but he is led to suspect that it was unsound because he was continually juggled out of the chance to test it. Our laboratory manuals too often give no definite indication of how far induction may go, and they seldom draw the line sharply between the knowledge obtainable from this, and that which must be obtained from some other source, if it is to be obtained at all. It seems to me that, unless the laboratory work is all reduced to plucking unopened buds, for mere practice in plucking, the practical work and the book must go hand in hand. The laboratory work is indispensable. It gives real knowledge of a kind no book can furnish. But the book must be employed also unless the instruction is to progress with the leisure and resources of original discovery (*cf. Heuristic Method*, p. 105).

e. *The Importance of keeping the Subject in Contact with Every-day Life:*[1] — There are two dangers into which the teacher of chemistry may fall. One is that of so circumscribing the view which he gives of the science that it is shut in by a high fence which precludes even a glimpse of the world beyond and its chemistry; the other is that of digression into various attractive and more or less familiar subjects, which may thus be allowed to interfere with the systematic teaching of the science. While avoiding the real danger of excessive digression, we must at all hazards save the subject from the former abuse by the judicious employment of legitimate means of illustration and vitalization.

This procedure is helpful to the instruction in the science itself. A strange subject, dealing entirely with foreign material, can never be interesting to the majority of pupils. Their natural craving for a tinge of human interest in everything is starved. Surely the study of a subject which is as intensely interesting historically and industrially as chemistry is, need never suffer from this limitation. On the other hand, we cannot possibly confine ourselves to common materials in attempting to teach the science. In speaking of oxygen, we immediately encounter barium peroxide, mercuric oxide, and potassium chlorate. If we were to attempt to avoid strange bodies like these, we should be bound to leave ourselves without means of systematically building up and rounding out the architecture of the science. We simply cannot summon it forth from a mass of information about cooking, agriculture, rusting, and photography by any legerdemain. The chemistry

Unfamiliar Materials.

[1] See pp. 68, 74, 129, 138, and 177.

of many of these things, and the experimental work involved in studying it, are too difficult for beginners. But when we speak of the three bodies just mentioned, for example, we can refer to Brin's oxygen process, to Priestley and his work, and to matches, so as to facilitate the introduction, on a friendly footing, of these barbarous materials, and so break down the shyness and reserve, if not distrust, with which new acquaintances will naturally be formed by the beginner.

Illustrations.

An illustration will help still further in showing what is meant. Suppose we state that aluminium is made by decomposition of the oxide by means of electricity, and that the equation is $2Al_2O_3 \rightarrow 4Al + 3O_2$. Bald teaching of this kind is not uncommon. Sometimes, in a misdirected attempt to animate the subject, the operation is explained in terms of the atomic theory. This, however, inevitably renders it more inanimate than before, and transfers it at once to a ghostly and unreal world. How much better to show the materials; to describe the plant, its location, and the water power it uses; to explain the process with its exciting details from the bath of molten cryolite to the blazing block of carbon at which the oxgen is liberated. The action of the electricity will be better appreciated if electrolysis of a dilute acid, or better still of cupric sulphate, is shown. *All* this need not be given, and certainly nothing like as much as this in connection with every chemical action. But once or twice in every lesson something is needed to revive the drooping imagination of the pupil, and give him a vivid stereoscopic view of chemistry as it is. Again, when we deal with the preparation of nitric acid, the method will be forgotten infallibly if some precautions are not taken. Reiteration is not a remedy! Why not contrast it with sulphuric acid, which cannot be made from the sulphates, found in great quantities in nature, for easily explained reasons. On the other hand nitric acid, although natural nitrates are expensive, cannot be made economically from the elements. In connection with its synthesis we have Cavendish's work on the investigation of the residual gas in air, then, and for long afterwards, supposed to be all nitrogen, and Lord Rayleigh's recent success in obtaining argon by use of Cavendish's principle (this, by the way, is an admirable illustration of the effect of removing one factor in a reversible action). Finally, the sources of the natural nitrates and their production under the influence of bacteria are available for lending colour and interest to the subject. It must be repeated that giving all of this would take time which cannot be spared, but something may be picked out in connection with almost every action in chemistry which will be helpful in making it comprehensible or memorable.

The employment of illustrations from things outside the laboratory also increases the usefulness of the instruction. The teacher in the secondary school, in view of the fact that his pupils are most of them receiving from him their sole preparation for life, has a certain responsibility in this direction which he cannot avoid. In a later section (p. 177) we shall have occasion to suggest a number of questions, for reply to most of which the basis at least is to be found in elementary chemistry. There is no use claiming that chemistry is a study of real things and not an artificial discipline, unless we show that it is so. It is not suggested that the applied chemistry should be taught, but only that its existence should be made plain in the most pointed manner. We need not and must not make too much of this aspect. Without there being any necessity for turning aside to teach domestic science, for example, and leaving chemistry altogether, there are illustrations of all sorts from various more or less everyday matters which must suggest themselves continually to the thoughtful teacher.

Still Other Illustrations.

Take, for example, oxidation. Aside from the hackneyed illustrations connected with rusting, life, and decay, the subject suggests the way in which blue clay becomes brown when exposed to the air through the change of the iron it contains from the ferrous to the ferric condition; the way in which paint 'dries' through absorption of oxygen by the solidifying oil; the way anglesite ($PbSO_4$) and cerussite ($PbCO_3$) are formed from galena (PbS) and are commonly found encrusting the veins of the latter; and the fact that deposits of ores of copper are mainly carbonate and oxide at the surface, and pass into sulphide as the exploration of greater depths proceeds. Reduction is illustrated by the various photographic developers and by the genesis of native copper from cuprite; adsorption by dyeing of cloth, the action of mordants, the formation of 'lakes,' and the effect of heating a glass bulb after it has just been evacuated to the point at which it begins to show the X-rays.[1] Reversible actions are illustrated by the storage battery; osmotic pressure by the root pressure in plants;[2] precipitation of calcium carbonate by the formation of coral and shells through the action of ammonium carbonate excreted by the organism inter-acting with the calcium sulphate in the sea-water; the subject of the lowering of vapour pressure in solutions by the spontaneous way in which impure table salt becomes moist, and ice is melted by contact with salt; dissociation of the true or reversible kind by lime burning; the displacement of metals by toning in photography; solution in the broader sense by reference to alloys and to gems like the sapphire or yellow diamond.

Other Benefits of Illustration.

The judicious use of this sort of illustrations involves not a loss but a saving of time, and it fixes points of real chemical value in the memory. The mention of things which are natively interesting in connection with other things which are not so is one of the best means of lending interest to facts that might otherwise seem dry. The practice is useful so long as it is employed with this end in view. It ceases to be so when the chemistry becomes secondary, and that which should be simply an illustration is dwelt upon to such an extent, or in such a way, that it displaces the chemical fact from the field of view entirely. Another advantage of this procedure is that it relates the subject continually to the physics, geology, and biology which the pupil may have already studied or may be

[1] Air adhering to the glass is liberated, and the X-rays vanish, to reappear only when the pump has been started again.

[2] Applications of the theory of solutions to physiology will be found in Jones' *Theory of Electrolytic Dissociation* (Macmillan), Chapters II. and IV.

about to study. It is at least as important that the teacher in the secondary school should give a general view of science, and of the relations of the sciences, as that he should give a sharply focused view of one.

Danger in the Abuse of Illustration.

Many chemists are conscientiously and systematically opposed to encouraging the introduction of anything which is in the least degree likely to divert the attention of the pupils from the science. They state, and probably with justice, that there is far too much so-called chemical instruction in the schools which is perverted into the teaching of odds and ends about various domestic and industrial applications of chemistry. I fear greatly, therefore, that what has just been said may be open to misconstruction, and may be taken as advocating, or in some way countenancing such a misuse of illustrative material. To be perfectly clear, let me say that the object of the references to every-day life will be defeated if they give occasion for long descriptions of these matters. It is a continual but brief reference to these matters in their chemical aspect, which shall show that the chemistry of the laboratory and the classroom is the same as that of the universe, and that there is such a thing as chemistry in the universe, that is suggested, and not such prolonged tours of superficial inspection of the chemical universe as will prevent that of the laboratory and classroom from taking definite form. The use of judgment of the sanest description is imperatively needed.

Need of Strict Scientific Correctness.

The facts cited for the purpose of illustration must be subjected to careful scrutiny before currency is given to them. The reader may recall the statement in a familiar work that "throwing water upon conflagrations results in the dissociation of the compound into the gases hydrogen and oxygen, which, in reuniting, add fury to the flames and increase the devastation." There are many popular notions about the scientific aspect of common things which contain fallacies more difficult to trace than is the absurdity in this. Then, too, the chemistry of many common things is but little known.

The contradictory statements about the reason of the harmful effects of the atmosphere of overcrowded rooms, for example, suggests that more knowledge of the subject is necessary before it can be admitted to the elementary classroom. Again, the chemistry of many familiar things is hard. Domestic science and agriculture are difficult to relate intelligibly to elementary chemistry. Soap-making can be explained, but the view of the facts connected with cooking and digestion afforded by the standpoint of the pupil in the elementary school must be too superficial and distant to make it suitable for incorporation in the elementary course. The treatment of these subjects must for the most part degenerate into the giving of mere miscellaneous information.

At the risk of repetition let it be said again that teaching the chemistry of every-day life is not the end of the course in the secondary school. Its object is the giving of discipline through a knowledge of chemistry in a broad but strictly scientific sense. Reference to the chemistry of matters of common knowledge is suggested simply as one means of attaining the main end of the course, by making the subject memorable, attractive, and digestible.

f. *Necessity for Unification of the Whole:* — Although the necessity for unification has been more than hinted at already, too much emphasis cannot be laid upon this point. In a multitude of details there is no wisdom. The mastery of the science consists not so much in steady accumulation of knowledge as in building up habits of observation and thought, and cultivating a chemical intelligence. No point in technique, observation, or generalization is ever past or disposed of. The pupil is slow to appreciate the universality of the ultimate constituents of chemical thought and work, and requires to have them brought to his attention again and again. The teaching of a science is a weaving process. The same warp runs through all, and, while the pattern develops and no strand is precisely like the preceding one, the result should be an harmonious development of the design as a whole.

Mastery of a Science Consists in Acquirement of Habits.

Each new fact is centrifugal in tendency and at first introduces a foreign element. The important thing is not the adding of new facts, but their utilization in the creation of that more recondite entity, the pupil's general grasp of the science. The quality of the result depends not on the amount or variety of the material, but on the perfection with which it has been assimilated.

g. *Some Misleading Words:* — One of the most fruitful sources of misconception lies in the ordinary phraseology of chemistry. Thus the word *strong* is very much overworked. It is used for *active* in reference to the chemical tendencies of oxygen, chlorine, certain acids, and so forth. It is used in connection with solutions when great *concentration* is meant. It is even used for *stable.* It will be found most satisfactory to use the proper one of these three terms in the classroom and to exclude the word strong altogether.

'Strong' Acids.

The words *stable* and, the converse, *unstable* are used in two ways. Thus sodium chloride and nitrogen iodide are respectively styled stable and unstable, in consequence of their behaviour when heated, and properly so. Phosphorus trichloride is exceedingly stable by this test, but is often spoken of as unstable because moist air decomposes it. In this sense everything might be considered unstable, since everything undergoes change when treated with a suitable chemical agent.

'Stable' Bodies.

The pupil eagerly learns a word and forgets the fact it was used to describe, and our words are often so badly chosen that afterwards, when the learner tries to reproduce the idea from the word, he makes the most egregious blunders. He will say that *water of crystallization,* for example, is water needed to make the particular substance, or all substances, take the crystalline form. He will say it is not chemically combined, because the word suggests a physical condition simply. It would be better to use the term hydrates exclusively (p. 96). At the end of the course, he will still say

'Water of Crystallization.'

that *oxidation* is combination with oxygen regardless of the many other phenomena it covers. He will say a *metal*[1] is a metallic-looking substance and call arsenic a metal, regardless of the fact that the term has a use of its own in chemistry. He will say that a *saturated solution* contains all that the liquid can hold of the dissolved body, and a *supersaturated* one more than it can hold (!), regardless of the fact that the terms have nothing to do with what the liquid can hold, but concern only what it can take up from a given sample. A solution of a certain concentration may be saturated, unsaturated, and supersatured all at the same time toward different forms of 'sodium sulphate.'[2] Our ideas in chemistry have so often been labelled wrongly that we must discredit the label while teaching the idea; our terms often obliterate and obscure the very distinctions they are intended to record. In examinations we must ask for illustrations, to make sure that the word has not been learned and its definition memorized without comprehension of what it really covers.

'Oxidation.' **'Metal.'**

'Saturation.' **'Supersaturation.'**

Some words are as yet without definite signification. Oxygen and ozone, rhombic and monoclinic sulphur, red and yellow phosphorus are called pairs of *allotropic* modifications. Yet the first pair, and probably the last, are chemically distinct substances, while the second is a pair of physical states of the same body, like ice and water.

'Allotropism.'

Absurdities, like the description of a metal as "brittle and ductile," are so common that we must heed the warning and make sure that even the definite terms are not being used as if they were a meaningless jargon.

h. *Some Common Fallacies:* — There are a few blunders, that have long since been recognised to be such by chemists, which still hang pertinaciously round elementary instruction. For example, the formation of hydrogen chloride by the action of sulphuric acid upon common salt does not prove the superior activity

[1] *Cf.* Tilden, *Hints on the Teaching of Elementary Chemistry*, 63–66.

[2] *Cf.* Walker, *Introduction to Physical Chemistry* (Macmillan), 50.

10

('strength') of sulphuric acid. Under other conditions hydrochloric acid displaces sulphuric acid from sodium bisulphate almost as completely.[1]

'Law of Precipitation.'

The so-called law of Berthollet in regard to the formation of precipitates, even in its least objectionable form of statement, is a half truth or less. So distinguished a chemist as the late Professor Cooke makes something like nonsense of it when he says,[2] "When materials are brought together in solution there is always a tendency to make such a transfer of their constituent parts as will produce insoluble compounds." It must be remembered that the principles of chemical equilibrium[3] alone, and not a tendency to produce insoluble substances, or, above all, any superior 'affinity' can explain this behaviour.

'Heat of Solution.'

The current explanations of the heat produced when substances like sulphuric acid and caustic potash are dissolved in water are almost all of doubtful correctness. The present condition of the science does not permit us confidently to offer any explanation and in elementary instruction it is safest to say so.

The early misconceptions produced by injudicious teaching in matters like those just cited are wonderfully lasting and can only be eradicated afterwards with great difficulty, if at all.

i. *The Grammar of Science:*— The ordinary treatises on the sciences omit all explanation of the fundamental conceptions of the scientific method. Indeed it would be fortunate if they also avoided introducing confusions of thought and blunders of the grossest kind when they touch the subject indirectly. The teacher of science must make up for this lack, and perhaps correct these misconceptions, by the study of some work dealing with the subject. We have space to mention two or three examples only.

[1] A. Smith, *Laboratory Outline*, p. 32.

[2] *Laboratory Practice*, p. 41.

[3] See Carnegie, *Law and Theory in Chemistry* (Longmans, Green & Co.) chapter VII., particularly p. 205.

Explanation,[1] its nature and correct use seem to be frequently misconceived. An explanation never attempts to state the reasons for, or causes of scientific facts. We can give no reason for chemical behaviour, nor do we regard it as proceeding from ultimate causes (in the sense of active originators which do things). An explanation is simply a *description* which relates a thing or idea to other more familiar things or ideas. In this way we explain the hastening of the evolution of hydrogen, when a little cupric sulphate is added, by reference to what we know about electric couples. An illustration fully worked out in detail has already been given in connection with the discussion of the use of the imagination (p. 131). The employment of terminology is not explanation. For example, to call an action 'catalytic,' classifies, but does not explain it. Note also that to call the *tendency to chemical action* 'affinity' simply substitutes the one word for the four. It also classifies, in so far as it distinguishes this tendency by name from cohesion and other forces. It most emphatically does not explain, for, instead of relating a fact to a closely allied but more familiar fact, it deliberately relates the simple fact to a complex set of entirely foreign ideas connected with the word affinity (namely, those of kinship, sympathy, and attraction), which are pure importations, and so the word confuses instead of explaining. Thinking that these ideas are germane to the thing itself, is, to apply a sentence of Wundt's, "one of those numerous self-deceptions which are no sooner stamped in verbal form than they forthwith thrust non-existent fictions into the place of reality." As a name for a thing and a means of classification, affinity is a good term; as an explanation, it is a failure, for, in the language of the schoolmen, at the best it is simply a case of explaining *idem per idem*, and at the worst of *ob-*

'Explanation' in Science.

'Affinity.'

[1] See W. K. Clifford. Aims and Instruments of Scientific Thought in his *Lectures and Essays*, particularly pp. 101–103 (2nd ed. 1886). Also Stallo, *Concepts and Theories of Modern Physics*, chapter VIII., particularly pp. 104–110.

scurum per obscurius, for it only adds greatly to the total to be explained.

The meaning of *law* in natural science seems far from being generally known. My own recollection as a student shows that for a long time I was in utter confusion as to the origin and meaning of the laws of physics and chemistry, because the meaning of the word 'law' had never become clear to me. Its usage was frequently so confusing that its significance could not be inferred. This experience I am convinced is not exceptional. The use of the word even in scientific books is so often incorrect that it would be astonishing if teachers were not in danger of misleading their pupils. The word is commonly accepted as applying to some dogma which requires no defence and must be accepted without question, or some belief, like that in our own existence as conscious beings, which is more easy to accept as an intuition than to support by argument. Another misuse of the term adds additional confusion. We find it stated that some gases do not *obey* Boyle's law; again, we learn that Boyle *discovered* this law, about a century and a half after Columbus discovered America. One writer speaks thus: "Nature . . . follows laws which are always operative under the same conditions. Vary the conditions of an experiment, and new laws are liable to intervene and change the result. *The essence of law is uniformity of action under like conditions.*" The italics are in the original.

'Law' in Natural Science.

Of the various senses in which the word is used, two only seem to be legitimate in science. In the narrower of these two senses,[1] a law is simply an exceedingly brief statement which embraces an immense range of separate facts. In a broader sense the term may be used of the uniform behaviour itself which is described in the statement of the law. It is therefore either the statement of a fact or it is a fact. In the latter sense it is pre-eminently a fact of the highest order. Thus the so-called law of falling bodies, for example, is either

[1] Karl Pearson, *Grammar of Science*, chapter III. See also The Duke of Argyle, *The Reign of Law*.

that which tells us in one brief statement all about the fall of every sort of material, whether it be a feather, a bullet, or the moon, and whether it falls towards the earth or some other celestial body, or it is the uniform behaviour itself of bodies left free to move under each other's influence. In the former and stricter sense this law was not discovered, however, for, unlike the New World, it did not exist previous to its 'discovery' and does not now exist objectively in nature. It was a statement invented or made from the comparison of multitudes of single observations. Laws in this sense are thus true only so long as they express successfully the facts with which they deal. When we discover by more careful observation that gases do not change their volume exactly in the inverse proportion of the pressure, it is not the gas which 'disobeys' the law, but the law which fails to express the facts exactly. Laws are not active agents; they do not 'operate' under any conditions, nor do they 'act' either uniformly or otherwise, and new laws do not 'intervene;' the humble law-maker has to change his law when he finds that the facts do not support it in its existing form. In the latter sense the law is the fact itself, the fashion of behaving. This was discovered, but it is not of the order of a mandate of some recognised authority, and the word 'disobey' is inapplicable. It is not supported by a police force, like legislative law, and therefore does not 'operate,' 'act,' or 'intervene,' even in the figurative sense in which these terms are used of the law of the land.

These unfortunate conventional modes of expression make it exceedingly desirable that the teacher should guard himself, first in his own mind, and then in the language he uses, most carefully against putting the idea of law in a wrong light. No single thing can do more than the misuse of this term to pervert completely the pupil's whole idea of scientific method and perspective, and to undo on a large scale what observation and induction are trying to do on a small one.

Of the matters coming up in elementary chemistry, the principles of definite and multiple proportions and combining weights are laws because they state facts, and facts of the widest bearing. On the other hand, Avogadro's statement, while it is frequently called a law, is not a fact of the same order as the others at all.[1] It is a part of the molecular theory of matter. If it be a fact, as it probably is, it is reached remotely by inference, and not directly by experiment.

'Cause' in Natural Science.

Misconception is particularly liable to occur in connection with the use of the word *cause*. When anything unusual or unfortunate occurs in every-day life, we immediately ask whose activity or negligence 'caused' the occurrence. We thus acquire the habit of looking for some active agent whose intervention is indispensable for the production of certain results. We have no justification for the use of the word cause in this sense in the scientific study of nature. We note occurrences, such as those connected with certain motions of bodies, and we sum up the nature of these occurrences in brief statements, of which the most condensed is known as the law of gravitation. Observation, however, leads us to the discovery of no active agent whose intervention brings the phenomena about. We do not know that the heavenly bodies will move to-morrow as they do to-day, or that iron will rust in the future under the same conditions as in the past. We regard it as probable in the highest degree that these occurrences will repeat themselves, but the relation is one of probability and not of necessity. We know the fact of each occurrence as a separate thing, and our general statement in regard to the occurrences has no power of enforcement for the future. In science, causes, in the sense of active agents which originate occurrences, do not exist. It is useless, therefore, to permit our minds to search for causes of this description.

Yet we have a tendency to furnish the link, which our habit of thought suggests as needful, by attaching the name of cause to something, and sometimes in doing this the term is grossly misused. For example, we sometimes hear the law of gravi-

[1] See Ostwald, *Outlines of General Chemistry* (Macmillan), chapter VII., in which the terms hypothesis and postulate are used.

tation spoken of as the cause of the behaviour of falling bodies. The mere statement which in one phrase epitomizes all the behaviour of falling bodies, should surely be the very last thing to which we should ascribe the power of bringing about occurrences which it simply describes. Even if we apply the term law to the uniform behaviour itself, this uniform behaviour may explain future occurrences, but it is not the cause of them.

Misuse of the Word 'Cause.'

A study of occurrences scientifically shows that they may be related in two ways. In the first place, certain phenomena are observed always to appear simultaneously. Occurrences connected in this way are described technically as being related by co-existence. The word cause cannot be employed in connection with any of them.

The other relation which may exist between phenomena is that of sequence. We pass hydrogen over a heated oxide and it is reduced, or we subject a match to friction and a chemical change occurs. The same phenomena are observed every time the same treatment is used. We do not know anything further than that this relation of sequence obtains. The result is a continual coincidence and nothing more. Its repetition on the next occasion is highly probable, but we perceive in the phenomena nothing which makes the consequence a necessity. In such cases we employ the word cause, and by this term we describe some one occurrence which always precedes some other. Pearson[1] says: "Cause, in this sense, is a stage in the routine of experience, and not one in a routine of inherent necessity." The term cause is misused when it is applied to phenomena which are related by co-existence, or when it is applied to gravity or affinity, which are facts simply, not causes at all.

Correct Use of the Word 'Cause.'

The teaching of science consists in establishing a point of view and not merely conveying a knowledge of facts. We should, therefore, avoid the misuse of words like cause and effect. When, through misuse of words like these, the rules of

[1] Karl Pearson, *Grammar of Science*, chapter IV.

the *Grammar of Science* are broken, obsession by false points of view ever afterwards distorts the victim's whole conception of the method of description and classification in which the study of science consists.[1]

'Matter' and 'Energy.'

Not to prolong our list, we may mention finally the words *matter* and *energy*. There is much difference of opinion as to the definition of these terms. It is certain, however, that some of the current statements are decidedly misleading. For example, it is sometimes said that the universe is made up of matter and energy. Again we learn that matter is the vehicle of energy. Still again we find the statement that energy is the cause[2] of change in matter. Each of these statements suggests an entirely different relation, and all are more or less misleading. Our whole knowledge of the universe is obtained by the study of our sense impressions. In describing these, we employ two conceptions. The idea of matter gives account of much that we perceive. Since, however, matter may be at rest or in motion, hot or cold, electrified or neutral, and the same specimen can change its state of motion, or of electrification, or its temperature, we separate these phases of the data of our experience, and employ a second conception, that of energy, for the purpose of describing them. Now we cannot logically describe one conception as the vehicle of another, any more than we can say that one axis in co-ordinate geometry is the vehicle of the other.

---

[1] It may be well to emphasize the fact that the discussion of topics like those handled in this section is by no means to come before the pupil. He could not understand their import and would only be confused. The teacher can convey correct ideas concerning 'law' and 'cause,' and 'energy' as readily as incorrect ones, if he is aware of the danger which lies in the employment of careless forms of expression, without openly discussing the terms themselves.

[2] In many actions the employment of some form of energy is antecedent to chemical change, in others the chemical change seems to be antecedent to the production of some manifestation of energy. There is evidently here no constant order of sequence. We have as frequent reason for saying that change is the cause of energy as that energy is the cause of change.

The conceptions must be kept independent, or they cannot subserve the purpose of describing the complex phenomena of experience. Matter and energy are not parts of the universe, but constituents of our mode of thinking about it and describing it. Philosophically they are best classified as conceptions of the mind and not things of an objective nature.[1]

---

[1] See Karl Pearson, *Grammar of Science*, chapter VII. and Stallo, who seems to use the terms force and energy as equivalent, *Concepts and Theories of Modern Physics*, chapter X. 149. *Cf.* James Ward, *Naturalism and Agnosticism*, Lecture VI., *passim*.

Karl Pearson's *Grammar of Science* gives an admirably clear account of many of the matters discussed in this section, and will be found to throw a flood of light upon the foundations of scientific thought. The whole contents of this book are bound to be a surprise to the student who has read only treatises on the individual sciences. The same subjects, in part, are treated with great clearness and philosophic insight in the first volume of James Ward's *Naturalism and Agnosticism*, chapters I.–VI. The teacher will also derive much benefit from reading Clifford's *Seeing and Thinking* and Mach's *Popular Scientific Lectures*, as well as Clifford's *Essays* and Stallo's *Concepts and Theories of Modern Physics*, to which reference has already been made.

CHAPTER VI

SOME CONSTITUENTS OF THE COURSE.

THE reader will have observed that a number of topics usually treated in chemistry have not yet received recognition in our discussion of the subject. Some of these are of very great importance, and some of them have a prominence in chemical instruction which is perhaps somewhat out of proportion to their intrinsic value. Our first section naturally treats of the atomic theory. This might perhaps have demanded a place in the chapter on "The Introduction of the Subject" at least as potently as the question of equations. Other topics of the same kind are valency, physical chemistry, and qualitative analysis. The order in which we take them is the order of their logical relation, rather than that of their importance as features in elementary instruction. If we had adopted the latter principle, the arrangement might have been different.

I. The Atomic Theory, its Nature and Place in Elementary Instruction.

What Facts call for this Theory.

a. *The Atomic Theory not a Fact:* — A theory is often formed by imagining a simple mechanical system, which would behave like some very complex subject of experiment in respect to certain clearly defined features in the phenomena presented by the latter, and to these features only of all the bewildering properties it may possess. The atomic theory is of this nature: it professes to explain certain features of chemical change. When we have found that each compound has a constant composition, there is no particular necessity for a theory to explain a fact in itself so simple. When, however, we learn that the compositions of several compounds containing the same constituents are related by the rule of multiple proportions, and that irregular quantities are never observed, we feel a certain satisfaction in thinking that if the constituents were done up in definite packets of uniform size, such a rule would be the inevitable consequence of the formation of several kinds of compounds. When later we reach the principle of combining weights (p. 75), and find that certain weights may be assigned to the elements, which, if they have been correctly chosen, will, in combination with the use of small multiples when necessary, express the weight in which the elements enter into all kinds of combinations, we feel an impulse to suppose that we are dealing with materials which are constructed, physically, like the interchangeable parts of a number of machines. The idea suggests itself that matter is so made that, if we could reach the ultimate parts of which a quantity of an element consists, by simply shutting our eyes and taking one or two pieces we should find that they would associate themselves with precision with other pieces of other elements to produce any number of different structures.

The Theory not Inevitably a Fact.

We should note, however, that, convincing as the theory seems, the facts which it explains do not in any sense constitute a proof that matter is really constructed as the theory demands. An illustration will make this clear. If we had never observed wheat or gold close at hand, and depended entirely upon the market quotations for our information about them, we should naturally infer that wheat was always done up in bushels, and gold in ounces. Yet the fact is that these substances have no such structure. They assume the forms of the bushel and the ounce only at the moment of measurement. So there might be imagined properties, at present unknown to us, which directed the quantitative selection of material for chemical change, and rejected the excess, without the existence of any permanent segregation into pieces of unalterable dimensions. The only bushels and ounces which we have are in the measuring apparatus and not in the material measured; so the only atomic weights may be in the properties controlling chemical combination, and not in the matter combining.[1]

b. *Its Limited Application:*—The atomic theory has been invented because of the difficulty we have in forming any mental image of the complex phenomena of chemical change as it takes place in large masses of matter. It furnishes a very fortunate suggestion of a mechanism which would exhibit some of these properties.

What the Theory does and does not Explain.

But the habit which chemists have acquired of speaking in terms of the atomic theory as if it described objective realities has obscured to some extent the fact that it does not attempt to account for everything in chemical change. It explains to us why 200 parts of mercury is the most convenient chemical unit, and describes the formation of mercuric iodide by the union of this amount with 254 parts of iodine in terms of the packet theory. It also accounts for the persistence of the masses of the interacting bodies, the only properties of the original materials which survive chemical change. In other words, it explains the quantitative relations. It makes no attempt, however, to picture to us the mechanism which would account for the disappearance of a shining, liquid, heavy, metallic substance, and another black, or perhaps we should say violet body, each with a definite set of physical properties, and the appearance of a scarlet solid with a totally different array of properties. That the *mere* placing of a particle of mercury very close to a particle of iodine, and in such a way that separation can still be effected at will, should lead to the production of a composite particle having properties markedly departing from the average of those of the constituents is inconceivable. For the explana-

---

[1] *Cf.* Stallo, *loc. cit.*, chapter VII. An entirely different mode of showing that the laws of combination might hold, even if the same identical little pieces of matter were not attached and detached in the course of chemical changes, is suggested by Karl Pearson (*loc. cit.*, chapter VII. § 6). *Cf.* also James Ward, *loc. cit.*, Lectures IV. and V. *passim.*

tion of the complete transformation actually observed, a much more complex theory would be needed (*cf.* foot-note to p. 75). To mention the atomic theory as furnishing such an explanation is to perpetrate a palpable absurdity. Even the youngest pupil must recognise the total inadequacy of the explanation, and only the submissive spirit produced by prolonged dogmatic instruction can prevent the criticism and rejection of so pretentious a claim. The use of the theory must therefore be confined at first to the explanation of the quantitative relations.

When valency is reached, the atomic theory finds once more useful application. Later still, in college work, when the relations of the constituents of a complex compound are described, the same theory becomes practically indispensable. Here, in the region of molecular constitution, it reaches the zenith of its success. But still, it is only so long as we hold it to the rôle of a theory and restrict its application to certain aspects of chemical change that it is of assistance. Nobody thinks that the molecules are in reality at all like our graphic formulæ. This application of the theory simply helps us to form a mental image of the generalized relations which subsist between the facts of chemical behaviour.

The Theory Indispensable in Later Stages.

The fact is, then, that in elementary chemistry the atomic theory attempts primarily to explain only the properties of combination by weight and volume, and it succeeds in this only because it leaves out of consideration the multitude of other less easily classified relations which make the actual phenomena so difficult to conceive and describe. When we thus relieve the atoms of the necessity of explaining the whole marvel of chemical change, they begin by being simply counters representing the combining weights; and, just as counters are not money, but have a numerical value, and assist in keeping account of transfers of money, so atoms may be regarded at first as primarily a fictitious medium of exchange, in terms of which we chronicle the account-keeping side of chemistry. It is only in the later application that the atoms become more and more concrete.[1]

c. *The Place of the Theory in Elementary Instruction*:—Having now cleared the way by pointing out the limitations of the atomic theory, we are prepared to take up the much-debated question in regard to the proper time for, and manner of its introduction in an elementary course.

Untimely Employment of the Theory.

One would fear being accused of uttering a platitude when stating, in the first place, that the use of a theory is to explain facts, and that it must, therefore, follow the facts to be explained, if it were not the most conspicuous feature of the situation that the atomic theory alone, of all the theories of science, seems to have gained a kind of prescriptive right to take precedence of the phenomena. The most cursory study of the text-books and the methods of the teachers of chemistry will show that this is the case to a predominating extent. And yet in the work which leads up to the noting of the qualitative characteristics of chemical change there is nothing which the atomic theory can explain or attempts to explain. Not only does it fail to explain the transmutation of iron and sulphur into ferrous sulphide; it is even flatly discredited by an unbiased consideration of the superficial features of the occurrence. Iron and sulphur, no matter how finely we divide them, and how closely we put them side by side, always remain iron and sulphur. The natural inference so far, therefore, is that the union must consist in something entirely different from a juxtaposition of minute fragments which retain their identities. If, therefore, we force the conventional so-called explanation on the pupil at this stage, he is bound to see that the doctrine is incapable of immediate assimilation with the experimental facts, that it is thus not of the nature of an explanation, and so he is driven to suppose that it is itself an inde-

[1] The most recent discussion of the atomic and molecular theories is in Professor Rücker's address before the British Association. The Structure of Matter: NATURE, LXIV. (1901), 470, or SCIENCE [N. S.], XIV. 425.

pendent fact. It seems to be unverifiable by experiment, so he infers that the science must include dogmatic teachings which are beyond verification. Long-formed habit makes the appreciation of oracular statements of this kind very keen, and so the dogma is at once given the place of honour in his estimation, and he starts his career as a student of the subject with a totally false view of the science and of the scientific method in general.

It is after the laws of multiple proportions and combining weights have been observed to hold true of chemical changes, that the opportunity for the explanation of these facts by the use of the atomic theory occurs. Its presentation at any time after this point is logically justifiable. It is very helpful in giving definite form to the conception of combining weights. It is not imperatively needed for the purpose of furnishing a concrete basis for thought and expression until Avogadro's hypothesis is introduced. It is exceedingly unlikely that, if the consequences of this hypothesis are to be developed at all, they can be made clear to a beginner in any other than the traditional manner. At this point the molecular and atomic theories will usually be employed frankly, although Professor Ostwald (in his *Outlines of Inorganic Chemistry*) has recently endeavoured to avoid them even at this stage.

Dangers to be Avoided.

The ever-present dangers seem to be those of forgetting that the atomic theory is a theory, and of permitting the limitations discussed above to slip out of view. There is also a tendency to treat the subject too realistically, and as if the behaviour of the atoms was the subject of direct observation. The latter tempts us, after we have observed the burning of magnesium, to say, without more ado, that the action takes place by the union of one atom of the metal with one atom of oxygen. If the pupil makes no measurement, we should tell him that weighing gives the proportion of the elements in this compound; that elaborate experimental investigation has assigned 24.36 and 16 as the most convenient combining weights of the elements concerned; that the proportion here turns out to be that of one combining weight of each element; and that hence, in terms of the atomic theory, one atom of each element is used. Without these essential links, all implied in the original statement, of course, but incapable of extraction from it by the inexperienced imagination of the beginner, the pupil cannot perceive the basis of our description of the action, or understand what it means. It would be better to avoid the term atom as much as possible in the every-day language of the classroom, and to substitute atomic weight[1] or combining weight for it. These terms at once recall the experimental method which is the true basis of every statement.[2]

Consequence of Misuse of the Theory.

I remember being present at a conference at which the subject of the present section was being discussed. The leader was inclined to favour explaining the very first chemical change as an operation involving atoms, and leading the pupil to think of everything that happened in terms of the atomic theory from the very start. I had just interposed some remarks expressing views similar to those defended here, and had wound up by pointing out that the science of chemistry dealt practically with the behaviour of gross matter and not with the vagaries of atoms, when a young man near me, who apparently had difficulty in restraining his impatience, burst out with the exclamation, "If chemistry is not all about atoms, what is it about?" He was a pupil of the leader of the conference and seemed never to have been led to realize, that, in the great department of knowledge which constitutes the science of chemistry, the complex processes of nature which it describes, the magnificent industries which it has founded and still guides, the services to the community which it renders in a thousand applications of analysis, and the multitude of distinct bodies and their rela-

[1] For the definition of the expression 'atomic weight' in terms of experiment, see footnote to p. 164.

[2] For the most satisfactory treatment of the atomic theory, see Hofmann, *Introduction to Modern Chemistry* (London, 1865), chapters X.-XII.; Ostwald, *Outlines of Inorganic Chemistry* (Macmillan), chapter VII.; Remsen, *Chemistry* (advanced course), chapter VI.

tionships with which it deals, there is scarcely ever any mention of the atomic theory, except as in so far as it may furnish an occasional figure of speech in their discussion.

The Art and Practice of Condensation applied to Chemistry.

The elementary chemistry of the classroom is, or should be obtained by a careful reduction of the whole subject to a smaller scale, and a careful elimination of the parts which in the microcosm have become inconspicuous details. This process cannot be carried out by any one who is not a chemist with a broad knowledge of the subject in the unreduced form. Too many writers of text-books have not the necessary qualifications for their task. In the process of eliminating all that can be left out in the most elementary work, and in arranging the remainder with a view to what is supposed to be a simple and pedagogically correct method of presentation for the use of the beginner, a good deal of distortion sometimes occurs, and the result of this severe banting process is in danger of becoming a mere phantom, if not a caricature of its former self. It is as difficult to recognise the ox in the beef extract, even when the latter is genuine, as the science itself in some of the epitomes which are placed at the service of the teacher. In theory, the elementary course in chemistry should represent the main features of the science *in petto*, and show it, as it were, viewed through the wrong end of an opera-glass. When the reduction has been carried out faithfully, the microcosm, if once more expanded, should reproduce in outline the image of the original. It is to be feared that were this re-enlargement attempted with much of the tabloid chemistry of the schoolroom, a monster of terrifying and most unnatural form would be hatched forth. It may be possible to introduce the atomic theory at an early stage without causing confusion, and to talk to beginners of atoms, when combining weights are intended, without obscurity, but I do not know any text-book in which this has been done successfully, and I could name many in which it has been attempted with disastrous results.

In conclusion, and at the risk of being accused of over-insistence, let us look at the matter from still another point of view. If we reflect upon what has been said in the opening section of the chapter on "The Introduction of the Subject," we may get additional light upon the question before us. We enumerated three distinct characteristics of chemistry, which, while they offer impediments at the beginning of the pupil's course, at the same time constitute the precise reasons for introducing a fresh study and a new kind of discipline. There was first the fact that it accustomed the student to knowledge-making with material objects and phenomena as the basis of this exercise. Atoms are not material objects whose presence and properties he can perceive by his senses, and thus do not furnish concrete material for the knowledge-making process. In the second place, the treatment of the subject was to be inductive, and was to start from a large body of facts, and, by the study of these, to lead to the elaboration of more general truths and more abstract conceptions. Now the atomic theory does not constitute a part of the fundamental data of the science. In the third place, the pupil was to be taught to rely upon his own powers of observation and inference, and to learn to discount the teachings of authority. If at the very outset we state that matter is composed of atoms, we ask him to accept on faith something which he cannot observe, and could never have found out for himself. Instead of asking him to exercise his own powers, we treat him to a dogma, and we cannot even render the dose palatable by furnishing convincing reasons for our belief, for we have none. Thus the teacher who dogmatically introduces the theory of atoms, violates every one of the conditions imposed by the nature of the subject, and proceeds treacherously to undermine the foundations of the very claims which have secured for the science admission to the curriculum.[1]

The Atomic Theory and the Claims of Chemistry to a Place in the Curriculum.

II. The Treatment of Valency.

When chemistry is treated in a mechanical fashion valency is a most important topic since chemical union is the subject of the

[1] See also pp. 79, 81, and 164, footnotes.

science and, in this view, is effected by the hooking together of atoms. The number of hooks which each possesses is naturally one of the first things we wish to know about. When, on the other hand, this burlesque mode of treatment is avoided, valency is still recognised as highly significant, but trouble is encountered, in trying to introduce it at any early stage, on account of the difficulty of explaining its experimental basis. I am inclined to think, however, that, if the way in which it arises out of experimental work is made sufficiently clear, there is no reason why valency should be quarantined as a thing which is unteachable without lapse into over-realistic atom mechanics. It is certainly not *primarily* 'the capacity[1] of an atom for holding other atoms in combination.' This is simply its interpretation according to the atomic theory, and ranges it along with the other facts of the science in harmony with the rest of this great conception.

Can Valency be Explained to Beginners.

Valency arises, of course, experimentally, after we have chosen the values for our combining weights which shall be elected to the proud position of atomic weights, and this choice, in its turn, comes after the explanation of Avogadro's hypothesis, the consequences of which determine the choice uniquely. An illustration will make clear exactly what is meant. Take, for example, the action of zinc upon hydrochloric acid ($Zn + 2HCl \rightarrow ZnCl_2 + H_2$). Before we have settled the atomic weight of zinc, we simply find that 32.5 grams of it displace 1 gram of hydrogen. After we have fixed the atomic weight of zinc as 65, that of hydrogen as 1, and that of chlorine as 35.5, by analysis and the application of Avogadro's hypothesis (*cf.* footnote, p. 164), our measurement tells us that 65 grams of zinc displaces 2 grams of hydrogen and combines with 71 grams of chlorine. One chemical unit of zinc therefore plays the part of two chemical units of hydro-

Experimental Origin of Valency.

[1] The word 'power,' frequently substituted for 'capacity,' in this phrase is inadmissible. Valency is not in any way a measure of the force with which atoms are held together, but only of the number of atoms that can be held by one.

gen and unites with two chemical units of chlorine. In consequence of this, we say that the chemical unit of zinc is bivalent. Valency is thus simply a consequence of the choice, amongst possible combining weights, of the final atomic weight. There is no new addition to the theory of the subject, no so-called 'theory of valency' involved.

Valency may be placed on a sound experimental basis, even before Avogadro's hypothesis has been discussed and atomic weights have been settled, by the device, which may have been used already for other purposes, of stating that the chemical unit has been settled on grounds to be discussed later, and that its value in the case of zinc is 65. The logical necessities of the case are satisfactorily met by plain indication of the experimental basis and perfectly clear delimitation [1] of any assumption which may have to be made.

---

[1] The nature of valency is so continually referred to as an exceedingly obscure subject that there must surely be some lack of clearness in the explanation commonly given. We may be pardoned therefore for a brief statement, in harmony with the above paragraph, of its nature and origin.

If gases had no properties which suggested Avogadro's hypothesis, how would the composition of chemical compounds be represented by formulæ, and what would be the values of the atomic weights (and therefore of the symbols)? We should find by experiment the composition of aluminium chloride to be *Al*Cl (*Al* = 9 instead of Al = 27), of zinc chloride to be *Zn*Cl (*Zn* = 32.5 instead of Zn = 65), of carbon tetrachloride to be *C*Cl (*C* = 3 instead of C = 12), of arsenious chloride to be *As*Cl (*As* = 25 instead of As = 75), and arsenic chloride to be *As*Cl (*As* = 15 instead of As = 75), and every chemical unit would have the same valency, that is, would be univalent. Where there were two different units used by one element, as in the case of arsenic, heavy type might be made use of for the larger equivalent, or formulæ like as_3Cl and as_5Cl (*as* = 5) might be employed.

Now what effect has the application of Avogadro's hypothesis upon this? We weigh equal volumes of the vapours of all the compounds of every element, so far as we can volatilize them, and record the weights of volumes equal to that occupied by two grams of hydrogen (or rather 32 grams of oxygen. If O = 16, the volume is 22.39 litres, — the gram-molecular volume) under the same conditions. On inspecting the quantities of the constituent elements found by analysis in these weights (now molecular weights) of all the compounds, we find that no compound of aluminium contains less than 27 grams of the element in the molecular

III. Use of the Results of Physico-Chemical Investigation.

Some of our works on physical chemistry have given so mathematical an aspect to this subject that the ordinary chemist has

---

weight, no compound of zinc less than 65, and no compound of arsenic less than 75, and that when the numbers are larger they are always integral multiples of these numbers. If we retain our old equivalents, but adapt our formulæ so that they represent molecular weights, we shall find Al_3 ($Al = 9$), or some multiple of this, in every formula of compounds containing aluminium and Zn_2 ($Zn = 32.5$) and C_4 ($C = 3$) in every formula of the compounds of zinc and carbon. The chlorides would therefore be Al_3Cl_3, Zn_2Cl_2, and C_4Cl_4. To avoid this complication, we write $\mathrm{Al} = Al_3$ ($= 27$), getting the formula $AlCl_3$ and we adjust the other cases similarly.

Thus in applying Avogadro's hypothesis, we have ourselves, in a manner, brought valency about. It is not a complication but a simplification of our way of representing chemical composition, particularly in the case of elements having more than one equivalent. It has the additional advantage that, when the atomic theory is introduced, it then suggests the consideration of different elements as having various capacities for holding chemical units of other elements, and leads to the use of graphic formulæ. The inestimable value of substituting a formula like CCl_4

$$\begin{array}{c} \mathrm{Cl} \\ | \\ \mathrm{Cl} - \mathrm{C} - \mathrm{Cl}, \\ | \\ \mathrm{Cl} \end{array}$$

for C_4Cl_4 is seen in the marvellous results which the study of organic compounds has yielded.

Experimentally, valency is the number of grams of hydrogen (or the multiple of 8 grams of oxygen) which are displaced by or combine with one gram-atomic weight of the element, when the gram-atomic weight is the least quantity found in the gram-molecular weights in all compounds of the element.

It may not be out of place to add that experimentally the atomic weight of an element is in general the smallest weight of the element which is found in the gram-molecule of any compound containing the element. The larger weights of the same element found in gram-molecules of many compounds are always integral multiples (in accordance with the principles of combining weights and multiple proportions) of this smallest weight. If, as may happen in rare cases, they are not integral multiples, a molecule containing one atom of the element has not been encountered, and the greatest common measure of the weights

been inclined to give it a wide berth. It is very certain that this side of physical chemistry has no prospect of admission to the school course, and may possibly not have been included in the training of the teacher. Its terrifying aspect, however, should not prevent us from recognising that many of its results are of the greatest interest and importance in connection with the study of the most elementary chemistry, and that we are also at liberty to help ourselves to the conclusions and leave the more theoretical portions untouched, if we so desire. Perhaps the most important use of these results is in the general influence which they will have on the teacher's point of view in all his instruction, if he keeps them prominently in mind, rather than in the extent to which they may be expressly imparted to the pupil. Many teachers, however, are even in favour of teaching some of the facts and theories of physical chemistry when occasion offers, and when there seems to be a prospect that they will really assist the pupil without putting upon him any fresh burden. There are certainly some parts of general chemistry in which this would seem to be possible.

Value to the Teacher, if not to the Pupil.

---

found is taken as the atomic weight. This smallest weight passes undivided from compound to compound as far as chemical experiment can discover. Whether it is incapable of subdivision is another question. In all probability it is. For one thing we cannot conceive of a piece of matter incapable of subdivision. Then, too, J. J. Thomson's work (see his very interesting *résumé* in an article on *Bodies Smaller than Atoms*, POPULAR SCIENCE MONTHLY, August, 1901) seems to have shown the actual existence under certain conditions of particles much smaller than the chemist's atomic quantities. The atomic weight is not that of the smallest particle that exists, but is simply the smallest subdivision of which there is chemical evidence. It will perhaps lead to clearness if the idea of a chemical atom being a round indivisible mass is given up, and there is substituted for it the idea of a bunch of smaller fragments which moves as a whole through chemical transformations.

Passing over from the atomic weight of experiment to the atom of theory, we consider the latter a chemical unit, and not necessarily a structural unit, certainly not the smallest particle that can be conceived (an absurd phrase). It is the smallest mass of a particular element of which we have chemical knowledge.

Most of the recent elementary books for secondary schools seem to pay some attention to the experimental side of osmotic pressure, and of freezing point and boiling point phenomena. These subjects form one of the links between chemistry and physics on the one hand, and between these two sciences and physiology on the other, and on account of the important strategic position which they occupy, it would seem that any attention they may receive can be fully justified.

Its Results are Capable of Application.

Osmotic Phenomena.

The decomposition of electrolytes by electricity is illustrated in several chemical experiments which are never omitted from any course. It seems not unnatural, therefore, that some explanation of the phenomenon should be given. The electrolysis of dilute sulphuric acid cannot be called a decomposition of water by electricity, unless the electrolysis of a solution of potassium nitrate is to be described in the same way. The statement leaves too much out of account. The fact, for example, that pure water and dry hydrogen chloride, or dry potassium nitrate, separately, are practically non-conductors, and are not affected by the current, shows that solution is something more than a mere mixture of the two. Perhaps carefully prepared demonstration experiments, with a limited amount of explanation, will be found to give more insight into this matter than long discussion could do, and effect a great saving in time. The drifting of the ions through the liquid, for example, can be shown in several ways.[1] The formation of the ions can be observed when cupric bromide (*cf.* Richard's *Harvard Outline of Admission Requirements*, 31), whose molecules are deep brown or black, is dissolved in water. As the

Electrolysis.

[1] It is so important that the teacher, at least, should be thoroughly familiar with this subject, that he should not fail to assist his own study of it by trying experiments for himself. A number of admirably devised experiments, most of which are easy to carry out, are fully described by A. A. Noyes in a most instructive paper (JOURNAL OF THE AMERICAN CHEMICAL SOCIETY, XXII (1900), 726: reprinted in the ZEIT. FÜR PHYSIKAL. CHEM. XXXVI. 1). Their performance will throw a flood of light on the whole subject for any one who is not already familiar with it.

liquid is diluted, the dissociation of the molecules leads to the change of the brown colour of the latter to the blue colour characteristic of the copper ions.

Two additional reasons for some attention to this subject readily occur to us. Its relation to physics, and the light which the chemical aspect of the matter throws on the knowledge the pupil has already gained in the physical laboratory, suggest the closer inter-relating of the chemical and physical views of the same phenomena by means of the theory which explains both. Then, too, the most startling recent improvements in the chemical industries have been in the direction of the employment of electricity for many purposes. Many manufactures, formerly carried out in other ways, are already, or are rapidly becoming, largely electrolytic. The preparation of aluminium, alkalies, bleaching agents, and chlorates are examples of this. We cannot now teach chemistry and avoid frequent mention of electrolytic operations, and we cannot well make these operations intelligible without some explanation of the theory.

Double decomposition is an old subject. So familiar is it that we do not always realize that it is after all rather remarkable. If we cause two salts to interact by heating them, the chances are that a most complex action takes place. When we mix their solutions the action is almost always simplicity itself. The solvent, far from being a mere bystander, has control of, and directs the action, so that it takes place rapidly and consists in a neat exchange of certain groups. Some explanation of this would certainly seem to be not out of place.[1]

Acids, Bases, and Salts.

The treatment of acids, bases, and salts is a difficult problem in chemistry. It is difficult even to define the terms (*cf.* Tilden, *Hints on the Teaching of Elementary Chemistry*, 66–68). To define a salt, for example, as a substance which is made in such and such a way, is to shirk the task of defining it altogether. We shall probably have to

[1] In this connection, however, see the work of Kahlenberg, JOURNAL OF PHYSICAL CHEMISTRY, V. (1901), 339–392, or abstract in NATURE, LXV. (1902), 305.

say that a salt is a substance which enters into double decomposition readily and whose solution is an electrolyte. An acid will then be a salt of hydrogen, and a base an hydroxyl salt. The theory of ionization throws a flood of light on the behaviour of these substances (see, for example, "neutralization," in any recent work on theoretical chemistry, such as Dobbin & Walker's *Chemical Theory for Beginners*, chapter XIX.).

The Battery.

In the battery we have again a collection of phenomena which are as interesting to the chemist as to the physicist. The duty of furnishing correlation between the different subjects of study, which is imperative in secondary school-work, suggests again the desirability of some use of the theory of ionization. This subject, too, is closely related to chemistry on account of the way in which the electro-motive force produced by various combinations forms a numerical measure of the intensity of chemical action taking place in the cell.[1] The order of the elements according to the electromotive series (p. 202) has many applications in chemistry.

Chemical Equilibrium.

Chemical equilibrium cannot be counted amongst the new developments of theoretical chemistry, for its beginnings were coeval with the discovery of oxygen, and its principles were clearly understood forty years ago or more. A strange reluctance, however, has been shown in regard to the recognition of its laws, both by investigators and instructors in chemistry. The blunders which have been made through failure to pay attention to them are only too familiar. The importance of these principles in explaining many of the commonest chemical changes may well awaken surprise at this strange neglect. The class of actions in which they find their chief application, and which must be misunder-

---

[1] Lüpke, in his *Elements of Electro-Chemistry* (Lippincott), describes and figures a large number of experiments illustrating this subject, and this feature of his book makes it most interesting and instructive to the reader who has not himself any experimental acquaintance with electro-chemical phenomena. A series of experiments in the same subject is described by Dr. Lash Miller in the JOURNAL OF PHYSICAL CHEMISTRY, IV. 599.

stood without recourse to them, namely, reversible actions, form a majority of the changes which the pupil encounters in elementary chemistry. The contrary statement, which one so often sees, is so palpably incorrect that one can but wonder what limitation the author was thinking of when he made it. The Brin method of preparing oxygen (*cf.* Newth, *Inorganic Chemistry*, 162) by means of barium peroxide and the chemical decomposition of mercuric oxide furnish examples at the very outset. Most of the chief changes at the beginning of the course are reversible, and actions of this class predominate more and more as the course progresses. Whether it is judicious to point this out to the pupil, or to discuss the consequences of the fact, may be matter for discussion. But there is no question that if the teacher is not familiar with this fact, and with the whole subject, he is likely to fall into egregious blunders, such as stating that sulphuric acid is stronger than nitric acid, and enunciating quasi-principles of a misleading kind, like the so-called 'principles' of precipitation and volatilization. Almost the only way to get clear ideas on this subject is to read the treatment of it in several different books, and to make some experiments illustrating the principle of chemical equilibrium for one's self. An admirable series of experiments, which, for the most part, may be performed with simple apparatus, is described by Dr. Lash Miller.[1]

Conclusions.

On the whole, perhaps, after this discussion of some of the bearing of physical chemistry upon elementary chemistry, it may be a question whether, in the average course, time will be found for anything more than a touch of the subject here and there. The conclusion is inevitable, however, that the teacher himself must be thoroughly familiar with the results of physical chemistry and its application to

[1] Lash Miller, Experiments Illustrating Chemical Equilibrium. JOURNAL OF THE AMERICAN CHEMICAL SOCIETY, XXII. (1900), 291. For a simple account of the subject, see Carnegie's *Law and Theory in Chemistry* (Longmans), chapter VII. *Cf.* references to chapters in Muir and Carnegie's *Practical Chemistry*, dealing with this subject, footnote to p. 216.

ordinary chemical phenomena, if the instruction he gives is to be thoroughly sound. A knowledge of this subject is one of the most indispensable parts of the equipment of the teacher of general chemistry, whether in school or college. Whatever ideas along this line may be communicated to the pupils will certainly be given in a very elementary fashion, and will be made thoroughly concrete by careful experimental illustration. In this way the keenest interest may be awakened. The theories will not be given for their own sake as separate topics, but strictly in explanation of phenomena that have been encountered, and only so far as they simplify and explain these phenomena. The teacher who has not had an opportunity of studying the subject, however, would do well to omit it from his instruction entirely, as there is nothing more dull and valueless than a theory which is not lucidly explained and adequately enforced by 'pat' application and experimental illustration.

IV. Shall Qualitative Analysis be Included, and if so in What Form?

REFERENCES.

Armstrong, H. E. Presidential Address before the Chemical Society of London. JOURNAL OF THE SOCIETY, LXV. (1894), 361. Reprinted in part, NATURE, L. 211.

Brace, Geo. M. An Article on Qualitative Analysis. THE SCIENCE TEACHER (New York), II. 173 (March, 1899).

There is perhaps no question in connection with chemistry as a study in the secondary school which has called forth opinions so sharply opposed to one another as this. The head of the department in one of our State Universities says: "If it is advisable to have lecture and recitation work, let it be at the end of the course, and have the two earlier terms taken up with qualitative analysis. . . . If the fortunate state of affairs exists that physics precedes the subject of chemistry, I think the lecture work as such is of little importance." This appears to suggest distinctly that even classroom work in general chemistry may be dispensed with, under certain conditions, and that the whole introductory course may consist in qualitative analysis. That, on the other hand, analysis should be excluded entirely is a view expressed with equal definiteness by a large number of committees and single authorities. What the argument in favour of making qualitative analysis the first and almost exclusive subject of instruction may be, it would be difficult to say. The statements supporting this view seem always to be of the nature of *obiter dicta.* The fact that analysis must at best give but a restricted knowledge of chemistry is so obvious that those who hold this opinion must do so either from lack of reflection, or on account of the influence of tradition, or perhaps they hold their belief simply *quia absurdum est.* Leaving this extreme view out of consideration, however, there are strong arguments presented by both parties, and we shall attempt to consider both sides. It is assumed, of course, that, in what follows, by qualitative analysis we mean the ordinary treatment which seems almost invariably to begin with the tests for the 'metals' and to confine itself for the most part to wet reactions.

a. *Arguments in Favour of Qualitative Analysis:* — There is, in the first place, the consideration that the school course may in its absence do little towards suggesting the variety of topics which is studied under the name of chemistry. The introduction of analysis tends to give a somewhat more complete view of the science by presenting another aspect of it.

It furnishes an excellent training in observation. Even if we admit that the phenomena of precipitation are not in themselves

Easy Observation.

exceedingly important, the facility with which they lend themselves to study by young students gives them a certain value. This argument is extremely instructive, for the satisfaction with which this sort of observation is made depends upon the limitations which the selection of phenomena useful for analysis has imposed. The possibilities are strictly limited in advance. This assimilates the work to that in Latin and mathematics, in which the pupil has become accustomed to similar guidance, and he welcomes the opportunity to study something which has similar characteristics.

The work exercises the reasoning powers and furnishes material for simple and sure inductions.

Analysis furnishes a review of certain parts of the subject which have been studied before, and thus assists in crystallizing the pupil's ideas.

The properties of the substances encountered may be studied in connection with the building-up of the analytical scheme, in such a way that genuine additions are made to the pupils' knowledge of chemistry and the bald enumeration of precipitates and colours is avoided.

A 'Practical' Subject.

Finally, it is well known that analysis is a subject capable of practical application. There is just a touch of something useful about it which awakens interest on the part of the pupils, and inclines parents and outsiders to approve its inclusion in the high school curriculum. Of course this rests on the mistaken notion so commonly held that chemistry consists in analysing things, and that the scientific chemist spends his time in examining groceries for adulterations and advising his neighbours in regard to the purity of their drinking waters. But it none the less creates a feeling of sympathy with the work of the school which has a value of its own. The interest of the pupil is natural, for, although the fact that silver chloride dissolves in ammonium hydroxide, while lead chloride does not, possesses no native interest whatever, the suggestion that this can be used for distinguishing the substances, or ascertaining the presence of lead or silver, gives it a powerful derived interest and awakens the attention of the scholar at once.

b. *Arguments against Qualitative Analysis:* — There may be some arguments against qualitative analysis in any and every form. There are certainly strong arguments against some types of instruction in it, especially those of a mechanical kind. We shall assume, however, for the most part, that the topic in its best and most rational form is alone to be considered, and that an inadequate basis in knowledge of the facts and theories of the science is the chief defect from which the instruction suffers.

To the first of the arguments just given it is replied that chemistry as a science has nothing to do with analysis. Analysis is an application of chemistry and an art practised with commercial ends in view, or a tool used in the prosecution of strictly scientific work. A little of it may enlarge the ideas of the pupil, but a greater amount must have a decidedly narrowing influence.

The 'Observation' Argument.

The strongest argument in favour of analysis is certainly that which points out the practice it gives in observation, guarded by such restrictions that complete and satisfactory study is within the powers of beginners. The inference from this, however, is that other parts of the subject should be systematized in the same way, in order that gaining a fuller and more complete knowledge of them may be put within the reach of the student of the elements. A systematic yet simple arrangement of certain facts has been made for the purpose of analysis; an equally simple and systematic working out of all aspects of the subject should be made for the purpose of instruction. The commercial impulse has caused the former development. The teacher does not yet seem to have learned the lesson which this plainly inculcates, and, instead of imitating the principle and applying it to the whole science, he has simply borrowed the fragment which the analyst has worked out for his own purposes, and saved himself much trouble by making up from it a large part of his course. There is an obvious risk that too much system may lead to mechanical work, but it certainly would seem that some improvement of this nature might be made without trespass on the zone of danger.

There is no question that analysis furnishes good exercise in reasoning, at least when it is taught in a rational manner and not by the mechanical use of tables. Mr. Brace, in the paper referred to at the head of this section, explains in an admirable way how the best use may be made of its aptitude for cultivating the inductive powers. As the study of the science progresses, a record may be made of the solubility or insolubility of each substance, and in the latter case of the colour and other properties of the precipitate. When the table has been completed, the pupils themselves can pick out a method of analysis. They note, for example, that three chlorides only are insoluble, and that therefore the addition of hydrochloric acid will lead to the recognition of the presence of one of the metals concerned. With but little assistance from the teacher or book they may be led to work out a complete system of analysis applicable to simple cases. There is, as we have said, no question of the rational exercise which this procedure gives, but we may be pardoned for asking what the subject is on which the reasoning is being expended. It is the solubility or insolubility of a large number of bodies. Now this question is of but slight interest to the chemist as a chemist, for so far we have not been able to explain most of the eccentricities of solubility. This sort of study puts calcium chloride and calcium fluoride in diametrically opposite classes. It assimilates arsenic and tin, nitrates and acetates.

The 'Exercise in Reasoning' Argument.

Suppose that after a complete study of this had been made, the class were next invited to distinguish sticky substances from brittle ones. Glue and phosphoric acid would adjust themselves in one group, while salt and napthalene (moth balls) would find themselves accommodated in another. Or suppose that substances were classified by their colours, iodine and potassium iodide would part company at once, and sulphur and sulphuric acid would know each other no more. In the same way, liquids might be distinguished from the solids. Mercury and copper would be no longer classed together as metals, and in this remarkable course ice and water would belong to different groups. Similarly, if the year was not yet exhausted, odours and specific gravities might afford opportunity for careful observation and nice discrimination. Devising systems of analysis based on these facts, and using them, would unquestionably furnish admirable exercise for the reasoning powers. The fact is that there are a great many ways of making hodgepodge of the chemical relations of substances by classification according to physical properties, but the study of chemistry itself cannot possibly be assisted by anything which makes havoc of chemical similarities, no matter how admirable the exercise it may furnish for the reasoning faculties.

The reasoning of analysis, that by which we distinguish salts of silver, lead, and mercury, for example, is exactly like that by which, in whist, we infer the contents of our neighbour's hand. Whist might with advantage be substituted for qualitative analysis, so far as training of the reasoning powers goes. The mere use of chemical bodies and chemical language in a study does not make it, *ipso facto*, chemistry. The valid reasons for studying qualitative analysis, without a foundation far broader than any fraction of a year's work can give, amount to showing that it gives good discipline, whether it contains any chemistry or not, and that, while being administered to the pupil, it may also be so liberally basted with a chemical sauce as to become a good imitation of a part of the science itself. Whist could be enriched by incorporating with it some study of the chemistry of cellulose and black and red ink, but this would not greatly advance its claims to inclusion in a course in the science intended to be systematic.

The argument that analysis may be employed to furnish a review of the previously acquired knowledge of the science is true to a limited extent only. There are other ways of effecting this review which cover the ground much more completely.

The final argument [1] that the analysis of the high school has

[1] There is still another consideration which affects some schools more than others, and some pupils in all schools. It concerns the pupils who are preparing for college. If some schools give much analysis, some little, and others none, some much general chemistry, some little, and others perhaps almost none, the various pupils from these schools cannot possibly be provided by any college with a fit course in continuation of their previous work. Only those who have devoted a year to general

a possible practical application, and teaches something useful, is plausible, but will hardly bear examination. In after-life the graduate of the high school hires an analyst if he wishes any work of this nature to be done. If he tries his own hand at it, he will quickly discover that the course given in the high school is of almost no assistance in the solution of any practical problem. The subject is far too difficult to be treated adequately in a part of an elementary course. The fact is that the pupil whose time has been taken up with analysis will only begin to discover after graduation the utter valuelessness of the supposedly practical instruction of which he has been the victim. It will not help him in the least to understand what is wrong with the battery that works his electric bell when it gets out of order, the way in which the dye adheres to his clothes, how mortar and plaster of Paris harden and cement sets, why his city dilutes its sewage with water to render it innocuous, why writing is more apt to fade than printing, how baking powder and yeast act, how photographs and fireworks are made, why 'tin plate' rusts so rapidly when it has started in one spot while galvanized iron does not, why hard water uses so much more soap than soft, whence and how iron, copper, and other metals are obtained, what becomes of all the soda and sulphuric acid that are manufactured, and a hundred other matters which are of interest to

Analysis as a 'Practical' Subject.

---

chemistry can be considered, for they alone can have anticipated any appreciable and easily definable portion of college work. The others may have anticipated, partially and feebly, portions of several college courses, and will perforce have to go without recognition of this anticipation.

A vigorous discussion of the subject of instruction in chemistry, with special reference to the place of qualitative analysis, by Professor Armstrong in the presidential address already mentioned (JOUR. CHEM. Soc. LXV. 361, or NATURE, L. 211), will be found particularly instructive. He seriously advises the postponement of the detailed study of qualitative analysis to the very end of the college or university course, after most of the other topics which usually succeed it have been taken up. See also Professor Perkin's vice-presidential address before the British Association (Report of the Association, 1900), reprinted in NATURE LXII. (1900), 479-480.

everybody, the basis for whose explanation might have been furnished in the high school course if the time had been better employed.

Conclusions. If such analysis as can be given prematurely furnishes no more chemistry than whist (aside from special efforts more or less artificially to graft chemical knowledge into its study), is no more or less capable of practical application than Latin, and positively tends to confuse what of settled order the previous work in general chemistry may have brought about, why cherish it on account of the opportunities for exercise in observation and reasoning it offers when a study of chemistry in a broader sense might give the same opportunities and be a support instead of a stumbling block in the acquisition of a knowledge of the science.

The choice of proper subjects of instruction places a great responsibility on the secondary school. The subjects, to speak figuratively, should be fit to form little bases of supplies for use in after-life, and the treatment should make plain the main lines of travel in each subject. If the chemical base is placed eccentrically with reference to the science as a whole, and the roads with which the pupil is familiarized turn out to have been by-ways instead of thoroughfares, the knowledge obtained will have been useless, if it does not even prove a burden, in the subsequent campaign of life.

In all this it will be understood that we are speaking of the trivial form of analysis, which is the only one it can assume in the hands of pupils not possessed of a broad and thorough training in general and theoretical chemistry extending over one or two years. After such a training, the questions of genuine chemical interest with which it is replete can be appreciated, and its study becomes in the highest sense a study of chemistry.

c. *Exercises in Identification:* — The satisfaction and apparent success attending the use of qualitative analysis in an elementary course show the need of some exercises which shall, if possible, possess similar characteristics, while confirming and amplifying the knowledge of the subject as a whole rather than harming it by the introduction of a side issue. We must have exercises furnishing opportunity for observation and reasoning, limited by suitable restrictions; exercises, if possible, with an object in view that will awaken the detective instincts of the pupil; and exercises that will at the same time review and enlarge the knowledge of chemistry. Can a study devoted more largely to chemical properties and less largely to physical ones than analysis be devised? Some teachers are beginning to think so. The attempt to commence with the wet reactions for the 'metals,' however, must be abandoned.

A Substitute for Analysis.

Suppose we give a pupil at some suitable opportunity some substance like red phosphorus or powdered charcoal, and ask him to discover what it is by employing any experimental means that occur to him, and to make a report showing conclusively that it is the substance he determines it to be and no other. In a simple case he may guess what the body is from its appearance, but the furnishing of a logical proof will nevertheless give him much exercise, and result perhaps in repeated rejections of the report. Suppose, next, that he is given powdered sodium carbonate, common salt, calcium nitrate, or sodium hydrogen sulphite for examination in the same way, his task being to discover the acid radical solely. Suppose again that later he gets more difficult cases, such as potassium chlorate, sodium iodate, ammonium iodide, potassium sulphide, or zinc acetate. Familiar substances like glycerine, alcohol, soap, and so forth may even be given if the chemistry of carbon compounds has received much attention. The particular substances employed must naturally depend upon the elements and compounds that have been studied. In each case he is instructed as before to discover what the substance is, and prove its identity conclusively by experiment. It will probably be necessary to give him a start by a little discussion of the experiments most likely to give the largest amount of information. It will soon be agreed that

The Method of Identification.

heating a substance by itself in a hard glass test-tube, and again, in the case of inorganic materials, in a common test-tube with concentrated sulphuric acid, are likely to furnish most quickly a considerable amount of information.

Effect of Heating.

When heated alone, few bodies fail to change in some way. Sublimation, melting, boiling, and evidence of decomposition are all significant and will be noted. If gases or vapours appear to come off, the pupil will have to reflect on proper means of recognising by their properties whether they consist of oxygen, carbon dioxide, water, sulphur dioxide, etc., or contain more than one of these. Then he must decide, when the gas has been identified, what substances could have furnished it, and, if possible, taking the other facts into consideration, decide what the unknown body was. The pupil should be warned to keep the residue from this experiment, since, for example, any one of many kinds of bodies may be the source of oxygen, and examination of the residue to determine what it really was may be the quickest way to limit the choice.

Action of Sulphuric Acid.

When the body is heated with concentrated sulphuric acid, similar observations will be made. Oxygen, carbon dioxide, sulphur dioxide, etc., may be recognised by specific properties. Dense clouds of fumes will usually indicate the halogen hydrides or nitric acid. Coloured solids like iodine and free sulphur may be seen. A few salts, such as phosphates and sulphates, will give no visible action. Of course everything noticed and the inferences drawn will be recorded. If these two experiments do not lead to a definite conclusion, the examination of the residue of the former will usually assist very materially. For example, it may itself be treated with concentrated sulphuric acid.[1]

Naturally the reports first presented will almost always be

---

[1] An account of the inferences which may be drawn from the results of these and similar dry way tests will be found in W. A. Noyes' *Qualitative Analysis* (1898), 54–56, in Valentin's *Qualitative Analysis* (1898), 213 and 231, and other similar works.

inconclusive. If a gas is given off which burns, it will be designated hydrogen. When the pupil is asked how he knew it was not hydrogen sulphide, he will probably not be able to recall any reason. On returning with a definite report, say, to the effect that it had no odour, he will then be confronted with the suggestion that it might have been carbon monoxide, and so a fresh investigation will be started. Or if he reports oxygen, and has not excluded the possibility of its being nitrous oxide, the fact should be pointed out. If he reports potassium chlorate, he is asked how he knew it was not the perchlorate or hypochlorite. Each of these questions at once appeals to him, his pride is stimulated, and he rushes off to renewed experiment and to investigation of his laboratory notes, or, if these fail, of the books provided for reference. Of course the success of the system depends upon the neighbours in the laboratory receiving different substances, and upon the alertness of the teacher in suggesting means of making the report more logical and conclusive. This sort of work invariably arouses the interest of the pupils to the highest pitch, and a single exercise in identification will teach them more about chemistry than they have learned in months of ordinary instruction, and they will be the first to draw attention to this fact. Of course it would have had neither interest, nor object, nor possibility of success without the previous drudgery.

Conduct of the Work.

This work is not capable of being arranged in a system so simple that its employment becomes mechanical. The range of observation is wider than in wet way tests, and chemical knowledge is demanded continually. That it furnishes exercise in observation and reasoning, however, does not require to be pointed out. That it takes advantage of the various qualities of human nature which hold the attention and even awaken enthusiasm, is one of its most conspicuous characteristics. That it must furnish a review of most of the chemical properties and modes of preparation of the non-metals and their compounds, and this in the most practical way, is evident. Above all, after one or two attempts at in-

Benefits of this Work.

vestigation of this kind, the pupil experiences a feeling of self-reliance and of ability really to do something with his chemistry, which is necessarily stronger and more gratifying than it would be in the use of a cut and dried system that left little room for variety of procedure and independent thought, like the common scheme of wet way analysis. It may be added that a pupil with but little training will be as successful in this work, provided, of course, no cases of exceptional difficulty are presented to him, as another who has been prematurely trained in formal analysis, and he will usually reach his results more quickly, and what is much better, will know precisely how he reached them. I have heard of students of analysis spending two days in looking for the acid radical in an alloy. Students of analysis are in danger of being slaves to the system and of using the whole of it every time. They are often like the guides to show places in Europe, who get on nicely if they are allowed to deliver their accustomed harangue without any appeal to their intelligence, but who are paralyzed if they allow themselves to be interrupted to answer a common-sense question.

Work of the kind we have described takes a good deal of time, but its extreme instructiveness far more than makes up for this. If any time remains at the end of the year, there is no reason why the process of identification should not be extended to the metallic part of the compounds given for examination.

V. The General Content.

It would be needless to attempt to suggest what other matters should make up the course in chemistry aside from the topics we have discussed. Facts, and many of them, are needed for the foundation of generalization and theory, and for the illustration of the chemistry of the important elements and compounds. The selection of facts for this purpose must differ in different institutions according to the capacities and needs of the pupils, the taste of the teacher,

Principles of Selection.

and a hundred other circumstances. As a general suggestion, however, a statement of the Committee on College Entrance Requirements may be recalled, as it is sufficiently important in this connection to justify quotation: "Facts incapable of correlation should be avoided as far as possible," and again: "The facts should be given as examples from various classes, and not as isolated things. Thus to speak of a 'standard method of preparing hydrogen,' whereby the action of zinc on hydrochloric acid is meant, shows narrow and infertile teaching. It should be shown that all acids are acted upon by a certain class of metals to produce hydrogen. Examples of both classes of metals should be given and the general principles derived. The reason for using zinc and hydrochloric acid in the laboratory can then be stated."

Outlines Suggested by Influential Bodies.

The outline of work for a secondary school course in chemistry, which is most fully worked out in detail, is that prepared by the Committee of Nine, to which repeated reference has been made already. They show what, in their estimation, is a proper content and order for presenting the subject, and by useful comments point out the relations of the theoretical topics to one another. They indicate also the way in which experimental illustrations should be distributed between the laboratory and classroom. It seems to me to be the best outline of what the American high school can and should be expected to do. Excellent scientific judgment and practical knowledge of the condition of the schools are shown at every point in the report.

The syllabus of the Examination Board of the Association of Colleges of the Middle States and Maryland is founded upon the report of the Committee on College Entrance Requirements. It is much less detailed than the above, but the scope of work which it indicates is practically the same. A useful list of laboratory experiments, which is well selected and representative, has been appended by the Examination Board.

The syllabus issued by the Regents of the University of the State of New York cannot be commended without reserve.

The laws of chemistry seem to receive too little consideration. The only one of the quantitative laws mentioned is that of definite proportions, yet they presently ask for a knowledge of the "theory of valency" which cannot be obtained without the study of Avogadro's hypothesis, and all that it implies. They include also the atomic theory, although there is nothing which calls for the use of this theory so long as the subjects of multiple proportions and of equivalent proportions are omitted. The laboratory work which is suggested is too brief. It contains also nothing on the chemistry of the metals and their compounds, and none of the fundamental principles of the subject are illustrated in it with the exception, strange to say, of the law of mass action. Even this is referred to, however, in so questionable a way (see Experiments 15 and 24), that it might better have been omitted. Of course everything depends on the use which a teacher may make of any outline, but this one certainly does little to discourage disjointed work and the neglect of the principles for accumulation of facts.

The detailed statement of the admission requirements to Harvard College contains much theory in proportion to the number of facts. It is exceedingly well put together and highly instructive. Probably the explanation of the emphasis upon theory may be found at least as much in the nature of the instruction in the college itself as in consideration for the needs of the pupils in the secondary school who may never go to college at all.

VI. The Selection of the Text-Book and Laboratory Manual.

This subject is closely connected with the last, and the choice of books must depend on so many circumstances that definite recommendations cannot be made. It may be useful here, however, to recall the points bearing on this question which have been discussed in the present volume, and to summarize them in their application to the choice of books. Naturally these statements are intended to apply to the average case only, and in common with all statements on difficult questions like this, must be subject to numerous exceptions.

In general, a book which gives a plain account of the subject without too much pedagogical pretence will be best suited to the use of the teacher who knows his subject. It is unnecessary to say that it should be accurate, and not only accurate in its statements, but it should present a view of the science as close to that occupied by the scientific chemist as is consistent with its elementary character. It should deal almost exclusively, so far as facts go, with the common elements, and a not too numerous selection of their prominent compounds. Works of reference can be used for amplifying the information which it gives.

The Text-Book.

The spirit of the book should be inductive; the laws should be carefully explained as summaries of facts which have been given and in close relation to them. Theories should likewise be closely related to facts and should follow them. The general treatment should be connected, logical, and lucid, and should make the unity, rather than the diversity, of the subject apparent.

The book should treat of general chemistry in a sound fashion and as a pure science. It should not, for example, be arranged as an introduction to analysis.

Formulæ should be kept in their proper places, and shown to be receptacles for the results of the study of each action. They should not in any sense appear to be themselves the end of study in chemistry. The way in which the facts are translated into formulæ, as a sort of language or shorthand for expressing them, should be explained clearly, that no misunderstanding may arise.

The outline of laboratory work or laboratory manual should fulfil the requirements which we have discussed in Chapter IV. It should plainly exhibit coherence in the study of each topic, or at least should be capable of yielding results in which this coherence may be brought out. The out-

The Laboratory Manual.

line of each experiment should be a thoroughly sufficient guide to the pupil, without being overburdened with detail, and without foretelling the result; the manner of presentation should encourage and assist thought; the selection should be judicious; and, above all, the principles of the science should be illustrated, as well as the facts.

CHAPTER VII

THE LABORATORY, EQUIPMENT AND ILLUSTRATIVE MATERIAL

REFERENCES.

Whitney, E. R. Equipment of Secondary School Laboratory. High School Bulletin No. 7. Albany, N. Y., The University of the State of New York. Pp. 665–675. The whole of the paper is valuable and has furnished many suggestions for this chapter.

Arey, A. L. Management of Laboratory Classes in Chemistry. *Ibid.* Pp. 676–678.

Gibson, James H. Selection and Care of Apparatus. High School Bulletin No. 1. Albany, N. Y., The University of the State of New York. Pp. 362–366.

Catalogues of dealers in laboratory supplies.

The attempt should not be made to give instruction in chemistry in any school which is not provided with a laboratory fairly well equipped for the purpose. It should certainly never be taught without laboratory work, and a poorly furnished laboratory means prodigious loss of time both to the pupil and to the teacher, and many difficulties in discipline and class management. If, however, the authorities insist upon the teaching of the subject when no laboratory exists, a strenuous effort should be made to provide some tentative arrangement of an inexpensive kind in order that this indispensable feature may not be entirely omitted.

I. Accommodations required.

First in order comes the laboratory itself, which should be large enough to hold the necessary furniture and provide plenty of space for the moving about which the work entails.[1] Close

[1] For full discussion of the arrangement of benches with reference to space and light, see Minot, Science [N. S.], XIII (1901), 412.

to this should be the store-room, which should not be made too small, if perfect order is to be kept amongst the material it contains. A large well lighted closet may perhaps serve this purpose in a small school. The classroom will probably be shared by other classes in physics and perhaps biology. A private room for the teacher is indispensable. A balance room, with shelves resting on brackets attached to the walls, is extremely desirable, as the distraction of attempting to weigh in a crowded laboratory interferes with care and exactness. The fumes of the laboratory also damage the instruments. A dark room for photographic work is a convenience.

Rooms.

It is needless to say that all the rooms should be well lighted, provided with sound-proof floors and partitions, and perfectly ventilated. Artificial ventilation by fans is the best, if it can be had. Aside from the ordinary heating arrangements, live steam for the production of distilled water and for the steam baths will be required. The rooms should be furnished with gas connections for lighting, and the tables and hoods with lighting or fuel gas for experimental work. Water should be provided on all the tables and in the hoods, and electrical connections are desirable.

One of the prime necessities is a willing and intelligent janitor, and the maintenance of perfect cleanliness through his efforts. Mr. E. R. Whitney puts this exceedingly well when he says: "The activities of the pupil are largely influenced by his surroundings. There is an intimacy between environment and conduct, and character is the outgrowth of conduct. A dirty, poorly lighted, inconvenient room, though designated as a laboratory, containing broken apparatus and dilapidated furniture, breeds slovenliness, disorder, and degradation. Bright, cheerful rooms, kept neat and tidy, supplied with good apparatus and inspiring pictures, will be a powerful aid toward the formation of high ideals and the arousing of noble aspirations."

The Janitor.

II. Laboratory Furniture.

a. *The Desks :* — These should be three feet in height with tops two feet wide, and the working places should be three feet six inches long. The total length of a desk will depend on the size of the room, but should in no case contain more than four or five places. The desks may be placed back to back in pairs. The tops may be made of some hard wood. Paraffin should be ironed into them. This furnishes almost perfect protection from damage. Plate glass (three-eighths of an inch thick, resting on rubber), slate, and tiling, are frequently employed. Perhaps the best material is a form of soapstone, made by the Alberene Stone Co., Chicago. It is indestructible. Glass apparatus is not more liable to breakage where it is used than when wood is employed. The tops of the tables should be clear, the shelves and connections being carried sufficiently high above them to make cleaning easy.

Desks.

As more than one section may occupy the room, the space under each place should be divided vertically. For two sections, two cupboards with one or two drawers above each will provide accommodation for the apparatus. If there are more sections than two, or if economy in equipment is desired, one cupboard containing the less breakable materials may be used by all the occupants in common, and from four to six drawers, occupying the other side, may be assigned to different individuals, according to the number of sections. Combination or ordinary locks or padlocks, to which the teacher carries a master key, will be necessary on all cupboards and drawers.

A shelf running down the centre of each double desk will hold a few reagents. Glass shelves with iron supports are admirable. They obscure the light less than wood, and are not harmed by acids. Six bottles containing the three common acids in diluted and concentrated form, and three containing solutions of sodium and ammonium hydroxides and sodium carbonate will usually suffice for general chemistry. The stoppers of the last three bottles should

Shelves.

be covered with paraffin, or rubber stoppers should be substituted for them. Ordinary glass stoppers continually become fast in the mouths of the bottles, which are broken in large numbers in the attempt to open them.

Gas and Water. The fuel gas may be led in a horizontal pipe under the shelf, or vertical pipes may rise to the surface of the table and terminate in a piece carrying four narrow exits for rubber tube connections. The sinks of alberene may either be placed in the corner between two adjacent and two opposite places, thus serving for all four pupils, or at the ends of the desks. In the latter case, an open trough, lined with lead or alberene, running down the centre under the shelf, is useful. Narrow exits for water placed over the trough furnish means for attaching condensers, and should be threaded for carrying water pumps. In any case the water faucet over the sink should be placed somewhat high, to prevent breakage of apparatus. It is exceedingly important that the exit of each sink should be provided with a cap perforated at the top, in order that at least an inch and a half of water may always remain standing in the sink. Thus strong acids are diluted before entering the lead pipe, and solids have an opportunity to settle. With this arrangement, ordinary lead and iron pipes will serve for the drainage of the laboratory, and will last for years. Without these caps the pipes are quickly eaten away, and become plugged up as well.

Other Fittings. If they can be accommodated, recesses for the stools and the waste jars,[1] of which there should be at least one to every four working places, may be provided under the desk. Sometimes economy in equipment may be effected with little loss in convenience by the use of projecting strips of wood perforated with two or three holes to serve as filter stands, and by fixing iron rods in the table to take the

---

[1] Buckets, "Buggy pails," made by the Indurated Fiber Co., Water Street, Chicago, are cheaper than stone-ware jars, look better, and last as long. The same firm makes "Keelers," circular, flat-bottomed, shallow vessels, which make excellent pneumatic troughs.

place of ring stands. Rods eighteen inches long and three-eighths of an inch in diameter will cost little, and will serve the purpose very well. The invention of some form of accommodation for the laboratory directions and note-books, which would not interfere with the use of the drawer or cupboard, or with the work being performed, would confer a boon upon the student in chemistry.

Burners.

When gas can be obtained, the Bunsen burner will naturally be used for heating. In the absence of this great convenience, a small apparatus for making gasoline gas is a substitute, which, however, is only moderately satisfactory. A convenient acetylene generator, and a special form of Bunsen burner for use with it, are made by J. B. Colt & Co., Boston. The alcohol lamp is feeble and expensive; it may be supplemented by the use of a gasoline blast [1] when higher temperatures are required. For many purposes, ordinary small kerosene stoves (see figure in Cooke, *ibid.*, 193) will be found useful in the absence of gas.

Hoods.

b. *The Hoods:* — For the performance of experiments involving the evolution of noxious vapours, well-ventilated hoods should be provided. One section of a hood will be required for every four or five workers. In some cases, the hoods are placed on the desks, in others, along the side of the room. Flues, in the lower openings of which gas jets can be lighted, will serve the purpose in the absence of better means of ventilation. If connection with a fan is possible, however, it should be made. The floors of the hoods should be clear in order that they may be easily cleaned. Gas and water connections are best placed below the floor of the hood, close to the front, and the rubber tubing for attachments is passed through a small hole opposite each stop-cock. One or two pipes for waste water should rise at the back of the hood and open flush with the surface. At least one sink should be ac-

[1] Convenient forms of this, which work satisfactorily, are listed and figured by the Chicago Laboratory Supply and Scale Company, 39 W. Randolph St.

commodated in a hood in order that ill-smelling liquids may be disposed of without discomfort to the occupants of the room. It should be fitted in the same way as the other sinks.

c. *The Side-Shelves:* — Conveniently accessible shelves should be placed against the wall for the accommodation of chemicals. The solids used in considerable quantities may be placed in large stoppered bottles (say one litre). For most of the chemicals smaller bottles (say 200 c.c.) will be sufficient. The liquids may be accommodated in half-litre bottles. The reagents should be carefully labelled and arranged in alphabetical order, according to some definite system, upon the shelves. The bottles and their places on the shelves should be numbered consecutively with asphalt paint. The labels should be painted with melted paraffin to prevent defacement. It will be found that the books of printed labels usually employ an unsystematic and often incorrect nomenclature, while the formulæ they give are frequently erroneous.

Reagents.

The solutions should always be made of a fixed concentration, which is marked plainly on the label. It is better to furnish ready-made solutions than to direct the student to make them, except in the case of special exercises, as the latter method gives uncertain results, and always entails great waste of materials. It should be noted that more than one solution of the same substance, differing in concentration, is sometimes required, and that in general the best concentrations are not the same as those used in qualitative analysis.

A list of the chemicals required can hardly be given, as it must vary somewhat with the work. Many text-books[1] furnish a list of materials needed. As regards special substances, it should be noted that red phosphorus can almost always be

[1] For example, Williams, *Elements*, 398; Perkin and Lean, 327; Newell, 381; Young, *Suggestions to Teachers*, 42; Cooley, *Laboratory Studies*, 139; Shepard, *Elements of Chemistry*, 343; Nicholson and Avery, *Laboratory Manual*, 125. Torrey, 475, and Arey, *Elementary Chemistry*, xi, give lists of apparatus only; the others, apparatus and chemicals as well.

employed as well as the yellow variety, and is much safer to handle. A solution of ferrous sulphate had better not be furnished, as it rapidly oxidizes. In place of solid ferrous sulphate, ammonium ferrous sulphate is preferable, as it keeps much better, and is not, therefore, so liable to give misleading results. A solution of this double salt containing a little sulphuric acid will keep for months without much oxidation (Noyes). A solution of tartaric acid should be made immediately before use, as moulds grow in it when the attempt is made to keep it on the shelf.

Miscellaneous Furniture.

d. *Other Laboratory Furniture:* — A cabinet containing drawers divided into compartments and filled with corks of various sizes is necessary. In the same place accommodation may be found for pliers, files, copper wire (thick and thin, say Nos. 16 and 22), and cork borers, all for general use.

A broad shelf attached to the wall, or a small table, furnished with gas connections and covered with a sheet of asbestos board, will serve for the blast lamp.

An ordinary table for readers and a book-shelf for the most necessary works of reference should not be omitted. The books will be used ten times more, if placed in the laboratory, than if they are to be found in a separate room only. The carrying of the books to the working places, however, should be forbidden, as otherwise they are sure to be damaged.

A blackboard, and, if the method of filing note-books in the laboratory is adopted, a shelf near the door, complete the furniture of the room.

III. Laboratory Equipment.

The Still.

For the general service of the laboratory, an apparatus for the preparation of distilled water will be needed. If the steam is sufficiently clean, any tinner can make an apparatus for its condensation at small cost. The worm should be made of tin pipe, as this metal is least affected by water and air. If steam is not available, the necessary boiler, preferably of copper, can be made to the order of the teacher. Many different varieties are on sale. A very compact one is made by Richards & Co., New York. It should be placed near a sink with running water, in order that the supply of cold water for condensing the steam may be readily attached.

Steam Baths. As each pupil is provided with but one burner, it is a great convenience to have a large steam bath for general use. In quantitative experiments some kind of steam bath is practically indispensable, and a large one is less expensive than many separate small ones. After trial of many forms, I have found that the following arrangement is perfectly effective, runs practically without any attention, and can never dry up, and so suffer damage from overheating. It consists of a rectangular box of sheet copper, four inches in depth, and of size according to the needs of the class and the space in which it is placed. The cover is carried upon feet projecting to the bottom of the box, and reaches to within a fraction of an inch of the top. It is perforated with openings one and three-fourths inches in diameter for the accommodation of the evaporating dishes. An ordinary iron pipe, one-half inch in internal diameter, closed at one end and perforated at intervals of an inch and one-half, rests diagonally on the bottom. At the open end it rises vertically and projects through a hole in one corner of the lid. At this point it is connected with the supply of steam. The outflow pipe for the accumulating water is attached so that its lower side is about an inch below the lid. This bath should be situated in one of the hoods, with the overflow discharging into one of the pipes provided for waste water. The whole apparatus can quickly be taken apart if cleaning is required. The steam connections should include a suitable valve to prevent the return of the water into the steam-pipe, if the supply of steam in the building should be shut off. Large sand baths are sometimes used in laboratories, but they become filthy from the spilling of material into them, and are difficult to clean. Luxuries, like electrically heated plates, are usually beyond the reach of the school laboratory. They are very convenient, however.

A pair of scales and weights for rough weighings will be needed. The platform variety, with a sliding weight which takes the place of the ordinary weights up to five grams (see figure in Newell, 12), is the best. The small weights up to five grams will infallibly be lost, probably during the very first exercise, if the common form of scales is used.

Scales, etc.

Foot bellows and a blast lamp, a barometer and a thermometer, are among the other necessary articles. In experiments which require large quantities of certain gases, much time is saved by the use of generators, or by furnishing the gases in the liquid or compressed form. Kipp's generators[1] are the most commonly used. Oxygen compressed in cylinders, and liquid sulphur dioxide in glass bottles resembling siphons in appearance are also obtainable.

The balances have already been discussed (p. 116). The chief source of trouble is the tendency which the pupils have to lose the small weights. If a sufficient number of sets of weights can be afforded, each pupil should receive one, and thus be held individually responsible for its return in complete form. If a more delicate balance is required for any special purpose, the Sartorius balance, No. 3 (the less highly finished pattern), will be found sufficiently delicate for all quantitative work.

IV. Apparatus and Chemicals and the Store-Room.

While the labour of managing this necessary accompaniment of the teaching of chemistry is exceedingly irksome, there is nothing which contributes to making the work successful more than a businesslike organization of the way in which the materials are handled. Each pupil should be furnished with a set of the apparatus of which

System for Distribution of Apparatus.

[1] For an inexpensive generator, modified from a design of Ostwald's, see AM. CHEM. JOUR., XXI. (1899), 70. Another form is described in SCHOOL SCIENCE, I. 88; a chlorine generator is described by Cornish, *loc. cit.*, 21. See also Peter's *Modern Chemistry*, 373.

he stands most commonly in need. Apparatus which he requires but seldom may be drawn from the store-room, and its prompt return should be exacted. In order that the pupil may feel his complete responsibility for the preservation of the materials, and may use them with care and lock them away systematically after work, it is a good plan to furnish him at the beginning with a printed or mimeographed list of the materials he has received. This may also show the price at which, when the apparatus is checked up, all articles missing, whether through having been broken, lent, or left lying about, will be charged. After comparison of the list with the apparatus, the former may be signed by the pupil and returned to the store-room, where it is kept as a receipt. At the end of the year, if everything is returned undamaged, no charge will be made.[1] There are, however, some pieces of apparatus, such as towels, files, and wire gauze which, if furnished at all, necessarily cannot be issued again after they have once been used. The Bunsen burner and clamp, also, usually become corroded during a year's use, and, if given out again uncleaned, any damage which may occur to them will be attributed to the previous user. It is a good plan, therefore, to have these articles completely renovated during the summer vacation in order that nothing but fresh apparatus may be given out.[2]

Lists and Blanks.

On the opposite page is shown, on a reduced scale, a sheet like that whose use has been suggested. The lower portion contains materials which are unreturnable, and are paid for at once if not furnished by the pupil himself. In filling this sheet, all the items which could possibly be

---

[1] Since the handling of money by the storekeeper is inconvenient, the best mode of securing payment for broken apparatus and for non-returnable materials is to require a deposit with the treasurer of the institution. In exchange for this the pupil receives a breakage ticket arranged so that the storekeeper may cut off the value represented by each transaction. Any balance is redeemed at the end of the year after the set of apparatus has been turned in and checked up.

[2] The Chicago Laboratory Supply & Scale Co. makes a special business of this renovation.

OUTFIT FOR GENERAL CHEMISTRY STUDENT.

*Laboratory*____ *Desk No.*____ *Locker No.*____ *Date*______190

Articles Returnable. — The articles on the following list are loaned to the student and they must be returned at the end of the course **clean, dry,** and **in good condition.** If any article is observed to be missing, broken, or in poor condition, report same to storekeeper immediately; and if any such article cannot be replaced at store-room, a line should be drawn through the name of article, for no allowance will be made after sheet is signed. Other supplies may be had as needed at store-room, where posted rules should be read. After checking this list carefully, sign your name in full and **return this sheet to store-room as soon as possible.**

Article	@	Article	@
1 Sand Bath	$ .12	1 Burette, 50 c.c.	$ .75
1 Pneumatic Trough	.50	1 Burette, 25 c.c.	.50
1 Tripod	.45	1 Graduated Cylinder, 100 c.c	.48
1 Iron Stand, small	.25	1 Flask, 125 c.c.	.10
3 Iron Rings, 3 sizes	.10	1 Flask, 250 c.c.	.14
1 Clamp Holder	.20	1 Flask, 500 c.c.	.18
1 Burette Clamp	.45	1 Distilling Flask, 30 c.c.	.07
1 Universal Clamp, small	.45	1 Dropping Funnel	.75
1 Mohr Pinch Clamp	.06	1 Funnel, 50 mm.	.06
1 Hofmann Screw Clamp	.20	1 Funnel, 75 mm.	.08
1 Crucible Tongs, iron	.35	1 Funnel, 100 mm.	.09
1 Deflagrating Spoon	.12	1 Funnel Stand	.45
1 Iron Crucible	.25	4 Squares of Glass, 5 × 5 cm.	.02
1 Bunsen Burner	.30	2 Side Neck Test Tubes	.05
1 Test Tube Holder	.08	2 Hard Glass Test Tubes	.06
1 Graduated Rule	.08	12 Test Tubes, 130 mm.	.01
1 Set Weights	1.50	12 Test Tubes, 180 mm.	.01¼
1 Horn Spatula	.10	1 Test Tube Rack	.45
1 Sponge	.20	1 Thermometer	1.00
1 Porcelain Boat	.12	1 Thistle Tube	.05
1 Porcelain Crucible, No. 0	.12	1 Hard Glass Tube, 10 inches	.10
1 Evaporating Dish, No. 00	.10	2 Marchand's $CaCl_2$ Tubes	.26
1 Evaporating Dish, No. 1.	.15	1 Watch Glass, 58 mm.	.03
1 Evaporating Dish, No. 3	.18	1 Watch Glass, 75 mm.	.06
1 Porcelain Mortar	.35	1 Watch Glass, 100 mm.	.10
1 Nest Beakers, without lip, Nos. 1–5 (1: .08, 2: .09, 3: .11, 4: .13, 5: .16)	.57	1 File, round	.08
		1 Cork Borer	.10
		1 Pair Shears	.25
2 Bottles, Narrow Mouth, 250 c.c.	.05	1 Clay Triangle	.05
4 Bottles, Wide Mouth, 250 c.c.	.05	Gas Tubing, 2 feet	.14
1 Bottle, 1000 c.c.	.15	1 Desk Key and Padlock, No —	.50
1 Rubber Stopper, 2-holed	.08	1 Locker Key and Padlock, No. —	.50

Local address______________

Home address______________ (Sign here)______________

____

Articles Non-returnable. — The articles on the following list must be paid for at once, with Chemical Laboratory breakage ticket, which must be obtained from ——. The student may return any of these articles when he takes the desk, if he already has them, or if he cares to get them elsewhere.

Article		Article	
1 File, triangular	$ .08	6 Inches Rubber Tubing, 3/16 inches	$ .02
1 Test Tube Brush	.07	5 Feet Glass Tubing	.05
1 Wire Gauze	.06	5 Feet Glass Rodding	.05
1 Towel	.10		
4 Sheets Filter Paper, No. 595, 47 × 54 cm.	.05	Total	.48

Date______________19 (Sign here)______________

furnished for use in general chemistry have been included. Many of the articles, while adding to the convenience of the worker, or of the teacher, as in the case of the weights, may be omitted, in order to save expense, without damage to the effectiveness of the work. Indeed, all the ordinary purposes of the secondary school course may be served by a list little more than half as long. Another much smaller form may be used when single articles are obtained from the store-room. These signed slips being filed, the location of the various pieces of apparatus is always readily ascertainable, and the responsibility of some pupil for their return is fixed.

The teacher, unless his class is an exceedingly small one, should not be burdened with the task of attending to the store-room. His place is in the laboratory, and his presence can never be dispensed with for a moment. The loss will be not so much to the teacher as to the efficiency of his work. An attendant for the few hours during which the laboratory is most in use will probably not be difficult to find.

Keys.

The store-room should contain a key-board for the keys of all desks, lockers, and rooms in the building, unless this is kept in the teacher's private room. It is convenient also to have in it some articles which are useful in preparing the materials used in the laboratory, such as a pair of tinner's shears, an iron mortar, and sieves with meshes of various sizes.

Sources of Apparatus.

In the matter of apparatus, the chief necessity is to have an ample supply of the smaller articles which are most used. Expensive pieces of apparatus can always wait until the stock of the other more necessary articles has reached a sufficient size. Most of our glass apparatus is made in Germany or Bohemia, but recently the manufacture of very good articles at reasonable prices has been begun by Whitall, Tatum & Co., of Philadelphia. In ordering apparatus for general chemistry, care should be taken to secure flasks with relatively wide mouths, and, if possible, to have the glass tubing and the stems of thistle tubes, etc., all of the same size, in order that constant boring of new corks may be avoided. In general, apparatus of thin glass should be preferred. Clamps, burners, and other hardware convenient in form and economical in price are made by the Chicago Laboratory Supply & Scale Co.[1]

Chemicals.

Unless very large quantities are needed, chemicals may be bought without disadvantage in this country. Baker & Adamson of Easton, Pennsylvania, make chemically pure articles for analytical work. Except in the case of the common acids, and a few materials which may be bought of a wholesale grocer, it is better to buy chemically pure materials for all purposes.

Care of Equipment.

The teacher of chemistry must be a person who is not simply interested in learning, but must be willing to give a good deal of time to the management of the material equipment of his department. The utmost system and order which circumstances permit should always be maintained. Articles of metal should be watched to see that they do not corrode, and proper measures should be taken for their protection if the fumes inseparable from the laboratory seem to have reached them.[2]

While, for the reasons stated at the opening of this chapter, a good equipment is exceedingly desirable, it should not be forgotten that much may be accomplished at very little expense when more means cannot be obtained. The teacher is more important than the laboratory, for a good teacher will know how to use and improve even a poor equipment. A good

---

[1] The following are amongst the prominent dealers who furnish apparatus and chemicals of all kinds: Eimer & Amend, New York; Richards & Co., New York; Queen & Co., Philadelphia; Henry Heil Chemical Co., St. Louis; Sargent & Co., Chicago; L. E. Knott Apparatus Co., Boston; Bausch & Lomb, Rochester. The catalogues of these and other firms, which are usually illustrated, give much information in regard to apparatus. These firms, as well as the Chicago Laboratory Supply & Scale Co., undertake duty-free importation of apparatus and chemicals.

[2] A valuable article on the care of apparatus by Inspector James H. Gibson is published by the University of the State of New York. High School Bulletin, No. 1, 362–366.

deal may be done with ordinary deal tables, a few bottles, and domestic substitutes for some apparatus. Some suggestions on this head will be found in Cooke's *Laboratory Practice*, 9 and 10.

V. Classroom and its Fittings.

This room will probably be used in common with teachers of other sciences. It should, therefore, be provided with a large desk on which there shall be room for the apparatus and specimens required for illustrating the work of more than one class. The top of the desk should be for the most part perfectly clear, in order that an uninterrupted view may be had of everything upon it. The gas and water supply may run along the under side of the edge next to the teacher, and small holes through the top opposite the various stop-cocks will furnish means of making connections through the use of rubber tubes. There should be a sink at one end of the table, at least, and several water faucets should be provided, one being used for the attachment of a water pump arranged so as to produce a vacuum or to furnish compressed air. Underneath the table convenient cupboards and drawers for the apparatus used in demonstrations will be required. For most purposes a pneumatic trough with glass sides, so that everything may be visible, is preferable to one lined with lead and sunk in the table. Shelving for acids and other chemicals should be placed in a convenient position. The hood, which should be connected with the ventilating system, may be placed behind the blackboard. The latter can be raised when the hood is in use. Openings in the table provided with a down draught and proper means of securing the removal of gases generated in the course of experiments are almost indispensable. In order that no time may be lost in preparing the apparatus for demonstrations, or in exhibiting the experiments, no conveniences which can be obtained should be omitted.

The Lecture-Table.

The seats may be placed on steps three feet in width, and each rising six or eight inches above the one in front of it.

Chairs with tablets facilitate the taking of notes. The windows should be provided with grooves and opaque shades in order that the room may be darkened when necessary, and some arrangement should be provided for the exhibition of charts. Most of the apparatus used in demonstrations will be the same as that employed in the laboratory, excepting that everything must be on a larger scale. The nature and use of the necessary apparatus is described and the apparatus itself is figured in the works of Newth and Benedict already mentioned (*cf.* p. 134). The more important special articles will be a number of cylinders of various sizes, chiefly used in experiments on gases, large test glasses, which are useful in showing precipitations, and Hofmann's apparatus for exhibiting the volumetric composition of various substances.[1] For class experiments with electricity, the storage battery is much more convenient than any other, and is in the end much cheaper, if means of charging it is available. Seven cells, with plates five and a half inches square, will be found sufficient for all ordinary experiments, and the whole of these will not always be employed. A stereopticon for projecting lantern slides and some experiments is very convenient. The growth of crystals, for example (ammonium oxalate is a good substance), is difficult to make clear without this means of exhibiting its progress.

Lecture Experiments.

VI. Illustrative Material.

Charts and collections of various kinds add much to the interest and value of the instruction. Articles which will serve the purpose just as well, however, may be made or picked up in various ways if a little trouble is taken. Charts, for example, made from good illustrations in books which are up to date, will be better in many cases than the

Charts.

---

[1] Simplified forms of this apparatus are sold by the L. E. Knott Apparatus Co., Boston. The best arrangement for demonstrating the equality of the volumes of hydrogen and chlorine given off in the electrolysis of hydrochloric acid is that devised by Lothar Meyer, and figured in the BERICHTE D. DEUTSCH. CHEM. GESELL., XXVII. (1894), 850.

antiquated and clumsy productions which are still sold. Frequently a pupil will be found who has talent for this kind of thing, and the collections of the school may be enriched without much expense through his assistance. They should be made on some material which will not be damaged by handling, such as stout tracing linen, or paper backed with linen.

A list of the elements with their atomic weights, compiled from the latest data by F. W. Clarke ($O = 16$), mounted on linen measuring 41″ × 61″, is sold by Eimer & Amend at $2. A similar chart of the periodic system, using the same data, measuring 58″ × 42″ is obtainable at the same price from the same firm. An excellent series of twenty-four *Chemical Lecture Charts* is published by Sampson, Low, Marston & Co., London. They are mounted on linen, measure 40″ × 30″ and cost about $13. They include the plant employed in many chemical industries, and some illustrations of theoretical matters, such as curves of solubility and the apparatus for measuring freezing-point depressions. A set of twelve charts illustrating industrial processes, mounted on linen and measuring 170 × 125 cm. is sold by Kaehler & Martini, Berlin, at 20.50 marks. Some other charts are mentioned in Eimer & Amend's catalogue.

The teacher will find it convenient, frequently, to prepare charts illustrating his own way of presenting the subject. A list of the metals in the order of the electro-motive force they show when arranged in conjunction with some other metal in a battery, known also as the order of solution tension, if hung in the classroom will find frequent application.[1] This order rep-

[1] This order is not the same as that of the old list of Berzelius, which, although hopelessly out of date, still appears in some works, but is the result of modern electro-chemical (*cf.* Le Blanc, *Electro-Chemistry*, chapter VI.) investigation. Omitting the less common metals, and arranging the others in order of decreasing solution tension, it is as follows:—

Alkali metals	Nickel	Bismuth
Alkaline-earth metals	Tin	Antimony
Magnesium	Lead	Mercury
Aluminium	Hydrogen (H)	Silver
Manganese	Copper	Platinum
Zinc	Arsenic	Gold
Iron		

resents at once the tendency of the element to form ions, the potential it acquires when placed in a solution of one of its salts, and its chemical activity, all in decreasing order. Thus each metal displaces those following it when placed in a solution of any salt. Note the place of hydrogen. The metals before it displace this gas from water and dilute acids, those following it do not. The latter are found free in nature, while the former, if they existed, would eventually become oxidized by replacing the hydrogen of water or weak acids. The stability of the oxides is exhibited also in some measure. As far as manganese, they are not completely reducible by hydrogen; after manganese, they are easily reducible. The stability of other compounds under the influence of heat follows approximately the same order.[1]

Portraits.

Portraits of chemists of historical prominence are attractive additions to the classroom, and frequently the remembrance of important matters in chemistry will be assisted by association with the appearance of the man. The NATURE series (London and New York, Macmillan) includes some very artistic likenesses, although they are perhaps too small for use in a large room. Kaehler & Martini publish a series of portraits including forty-eight scientific men, with biographical text by Siebert (size 29 × 39 cm.). Portraits of Hofmann and Victor Meyer suitable for framing may be obtained of the Pharmaceutical Review Publishing Co., Milwaukee.

Photographs.

In this connection it may be pointed out that photographs made by the teacher, or some friend, from actual objects of chemical interest, such as parts of chemical factories, may be enlarged or made into lantern slides and furnish a valuable means of illustrating many things. Many of the charts and illustrations in books are so diagrammatic in their nature, not to say so completely out of date in many cases, that they give an exceedingly inadequate impression of the chemical

---

[1] A chronological chart exhibiting certain historical data is given by Tilden (*Hints on the Teaching of Elementary Chemistry*, 42–43) and may be found useful.

industries as they are. Authentic representations of the real thing, therefore, have great value in holding before the mind of the pupil the fact that chemistry is on one side a great industrial reality. They also assist in keeping the subject in touch with matters of every-day life which may, in many cases, have more or less close connection with the future business of some of the pupils. It is needless to say that visiting factories, so far as they are accessible, will be of the highest value. The managers are usually willing to allow a teacher to take his class to visit their plants, and will usually furnish a conductor more or less capable of answering questions intelligently and explaining the machinery and processes used.

Minerals. The illustration of classroom work by exhibition of specimens of minerals is also desirable. This strengthens the link connecting chemistry both with geology and with industry. Good specimens can frequently be found by the teacher or obtained as gifts. They may also be purchased from many dealers. Large cabinet specimens are not required. The most instructive specimens to purchase, when limited means only are available, are single crystals showing common and typical forms of the various substances. These may be obtained in almost all cases for ten or fifteen cents each. A set of typical minerals fulfilling these requirements need not be extensive.[1]

---

[1] List of 36 *Minerals* which furnish good crystals, are important ores, or are conspicuous constitutents of rocks [cr. (= crystal) and mass. (= massive) indicate the best forms for our purpose].

Copper (ramifying)
Arsenic (scales)
Sulphur (cr.)
Halite (NaCl, cr.)
Fluorite (CaF_2, cr.)
Cryolite (3NaF, AlF_3, mass.)
Calcite ($CaCO_3$, cr.)
Dolomite ($CaMg(CO_3)_2$, cr.)
Siderite ($FeCO_3$, cr.)
Arragonite ($CaCO_3$, cr.)
Malachite ($CuCO_3$, basic and hydrated, mass.)
Malachite (pseudom. from cuprite, cr.)
Selenite (Gypsum, $CaSO_4$, $2H_2O$, cr.)
Barite ($BaSO_4$, cr.)
Corundum (Al_2O_3, cr.)
Specularite (Fe_2O_3, cr.)
Haematite (Fe_2O_3, mass.)
Limonite (Fe_2O_3, hydrated. Pseudom. from pyrite, cr.)
Pyrolusite or manganite (MnO_2, hydrated, mass.)
Magnetite (Fe_3O_4, cr.)

One of the subjects strangely neglected both in schools and colleges in this country is the study of crystals. Their treatment here is in marked contrast to that in Germany, where a pretty extensive knowledge of crystallography is required of teachers in secondary schools, and a large part of the work in science[1] deals with the study of crystalline forms geometrically and with physical crystallography. It is not suggested that the time available in the secondary school course is likely to permit the introduction of much of this subject. Some trouble should be taken, however, to give the pupils an intelligent knowledge of how crystals grow, and of some of the common forms. The chemist depends very largely on the making of crystals for purification, and on the form of them for identification in his work, and both of these features appear in elementary chemistry, whether particular attention is paid to them or not. The pupils will always take great interest in growing large crystals for themselves, and will learn much from the exercise. Common alum and chrome-alum give beautiful octahedra; nickel sulphate ($NiSO_4$, $6H_2O$), illustrates the square prismatic system; cupric sulphate ($CuSO_4$, $5H_2O$), the asymmetric; double potassium cupric sulphate, made by mixing the two salts in equi-molecular proportions, the monoclinic, etc. Models made of wood or of cardboard to show the common forms on a large scale may be purchased or made very readily.

Crystals.

Chromite ($FeCr_2O_4$, cr.)
Cassiterite (SnO_2, cr.)
Quartz (SiO_2, cr.)
Sphalerite (ZnS, cr.)
Stibnite (Sb_2S_3, cr. Japan)
Cinnabar (HgS, mass.)
Galena (PbS, cr.)
Pyrite (FeS_2, cr.)
Zircon (cr.)
Garnet (cr.)
Apatite (cr. Ontario)
Cyanite (mass. Illustrates two hardnesses)
Analcite (cr.)
Hornblende (cr.)
Orthoclase (cr.)
Topaz (cr. Japan)

Prominent dealers in minerals are G. L. English & Co., New York; E. A. Foote, Philadelphia; Roy Hopping, New York; and Ward, Rochester.

[1] See Russell, *German Higher Schools*, chapter XVII., particularly p. 364.

VII. The Teacher's Private Room.

A private work-room should be provided for the teacher in order that he may have a place in which to pursue his own work undisturbed. He may there try new experiments for demonstrations, and perhaps devise better means of illustrating important points in chemistry for himself. He will also thus be enabled to continue his study of the subject by experimental work, for no one can afford simply to rest upon what he knows: such a course must really involve retrogression. If his appliauces and time permit, and his previous training has been sufficient, this room will furnish opportunity for carrying on research in some direction.

The room, like the laboratory, should have connections for gas, water, and electricity, and, in addition to the usual apparatus, should perhaps be furnished with a bench fitted with a small anvil and vice, and provided with a few tools.

CHAPTER VIII

THE TEACHER, HIS PREPARATION AND DEVELOPMENT

Nichols, E. L. Paper on the Training of Science Teachers for Secondary Schools, and discussion thereon. High School Bulletin No. 7. Albany, N. Y., The University of the State of the New York. 1900. Pp. 630–650.

Russell, J. E. German Higher Schools. London and New York, Longmans, Green & Co. 1899. Chapters XVIII. and XIX.

Bolton, F. E. The Secondary School System of Germany. New York, D. Appleton Co. London, Edward Arnold. 1900. Chapter II.

THIS chapter naturally divides itself into three parts which treat of the training of the teacher, the best means for securing his continued development, and the literature which will be most useful in connection with the latter. It would be useless to discuss the qualities of sympathy, tact, alertness, force of character, etc., which are indispensable in the teacher of chemistry as in the teacher of any other subject. These depend largely on the natural aptitude of the aspirant to the profession of teaching. It is rather the strictly professional part of the preparation of the teacher which primarily concerns us.

I. The Training of the Teacher.

The indispensable acquisition of the teacher is a well-rounded and sound knowledge of the subject. Nothing can possibly make up for the absence of a preparation which will give this. It is to be feared that the attempt is often made to teach chemistry without this prerequisite. Often, as we have already remarked, a teacher who is conscious of incompetence is required by the principal of his school to teach this subject, simply because its representation in the curriculum is desired. Often the student in a college whose curriculum is of the old stamp does not discover until late in his course the natural bent which he may possess towards physical science. He may thus, while lacking the proper preparation, find that his taste leads him in the direction of science, if his inclination or circumstances induce him to become a teacher at all. Often, too, the college student may have pursued the study of chemistry pretty extensively during his course, but the nature of the instruction may have been such that, in spite of his acquaintance with many phases of the subject, he is little better prepared to teach it than the members of the two other classes. For these and many other reasons, it is to be feared that the teachers of chemistry in our secondary schools, as a class, are not so thoroughly fitted for their work as they should be. Yet, as Professor Bennett says, we cannot "pass judgment on the mass of the incompetent. They are almost without exception men and women of character, of serious and earnest purposes, and faithful even to the detriment of their health in the performance of their tasks. They are, nevertheless, endeavouring to achieve the impossible, — to perform a work involving the employment of large resources without ever having secured the necessary preparation."

Knowledge of General Chemistry the Prime Essential.

The first constituent of this necessary knowledge of chemistry is general chemistry. If we ask what the second ingredient must be, we should be compelled to say again general chemistry, and the same answer must be given at every repetition of the question. It is a knowledge of the science as a whole and not of any special section of it which will count in elementary instruction. Only in so far as other branches may contribute to this knowledge are they to be considered a specially desirable part of the training of the teacher of elementary chemistry. It is a delusion to suppose that general chemistry can be disposed of in three months, and that the next thing to be done is to study qualitative analysis. A whole year of general chemistry will not confer anything like sufficient knowledge of the subject for our purpose.

Taking the matter in detail, we require first an introductory course. This must be thoroughly sound and fairly extensive. How rare courses possessing these characteristics are, only those who have studied the instruction in many institutions know. General chemistry cannot be taught by a public analyst in his spare moments, by a physician with limited professional practice, or by a "sticket minister" with a taste for science. The instructor must be a man himself engaged in productive chemical work and thoroughly abreast of the times. He must be, so to speak, a self-luminous body, for, the more he plays the part of a reflector or a refractor of borrowed information, the less truly will the image represent the nature of the science. The introduction should be a year in length, it should be accompanied by much laboratory work, and its whole scope should be much greater than that of the corresponding course in the secondary school.

First Year of Preparation.

Beyond this, the knowledge of the subject must be deepened in various directions. More acquaintance with the ordinary facts of the science, more knowledge of theoretical and physical chemistry by study and by practical work, more ability to handle the literature of the subject, and a far broader grasp of the ramifications of the science in the directions of industry, agriculture, geology, physiology, and hygiene are needed. And all this will be valueless if the theory, the literature, and the applications are not treated in a thoroughly modern manner.

It might be suggested, as a tentative plan, that the second year, following general chemistry, should begin with a study in classroom and laboratory of such topics as chemical equilibrium, the methods of measuring chemical affinity, and the boiling point, freezing point, electrolysis, and other properties of solutions which are of such importance in the chemistry of qualitative analysis. Without these preliminaries, the last-named subject can contribute nothing worth mentioning to the student's knowledge of general chemistry. This might be followed by an elementary study of quali-

Advanced General Chemistry.

tative analysis itself, care being taken to use the light which the recent study of solutions has thrown upon their chemical nature in explaining the *rationale* of the processes used, and in general so to employ the subject as to deepen and broaden the pupil's knowledge of general chemistry as far as possible.[1] Following this, exercises in the determination of molecular weights, the measurement of equivalents and combining weights, in which the refinements of quantitative analysis are employed and the results are used for working out atomic weights, will probably occupy the remainder of the year.[2] Throughout the course,

[1] Ostwald's *Scientific Foundations of Analytical Chemistry* (Macmillan) shows in detail how the operations of analysis may be rationalized.

[2] Experience shows that students gain but a feeble grasp on the science until they have done some exact quantitative work. It is preferable on many grounds even to begin the second year with three months of quantitative analysis, to follow this with theoretical chemistry, and to place elementary qualitative analysis last. Of course the benefit derived from the reversal of the ordinary arrangement will depend entirely on the method and spirit of the instruction.

Quantitative analysis should be used to train the prospective teacher in, and make him familiar with that accuracy of work and refinement of method, which are not only characteristic of the subject-matter of the science, but which also, in some shape or other, are the ultimate basis of all advance in the knowledge of chemistry by experimental methods. Such training is not only necessary if the teacher is to add to our knowledge of chemistry (see p. 216), but is equally indispensable if he is to understand, without hiatus or distortion, how our knowledge has been developed (see p. 77). To achieve these ends the quantitative analysis should not only deal with the separation and determination of a certain number of bodies, but should develop as far as possible a sense of the ultimate exactness and rigidity of the proofs of those theories to which in the previous work in general chemistry (and, when the old order was followed, qualitative analysis) constant reference has been made (see p. 72).

At the same time the previously acquired knowledge of chemistry in the broader view (general chemistry, as we have called it) should be used and increased as far as possible, e.g., by exact determinations of combining weights, by testing the law of the conservation of mass, and by applying the laws of chemical equilibrium to the methods used. Certain phases of general chemistry can be considered profitably at somewhat greater length at this stage, e.g., isomorphous mixtures, the preparation of chemically pure substances, and the growth of crystals of difficultly soluble salts such as barium sulphate.

Finally, in quantitative analysis students should take up some prob-

the reading in various works of reference, in selected original papers, and along historical lines, should be arranged with great skill, so as, on the one hand, to strengthen the pupil's grasp on the topics taken up in the laboratory, and, on the other, to fill out the gaps between these topics, and render the whole study more symmetrical. Indispensable as extensive reading is at this stage, it is almost wholly neglected in most institutions at the present day. It is left to the initiative of the pupils who have special interest in the subject, precisely at the time when guidance and stimulus from the teacher are most needed.

The difficulty in endeavouring to give a course like the above is that no text-books or laboratory outlines of the sort which would harmonize with this ideal are available, excepting perhaps in organic chemistry.

Organic Chemistry.

Inorganic Preparations.

Reading.

The third year of work will contain organic chemistry and inorganic preparations on the lines of Lengfeld's *Inorganic Preparations*[1] (Macmillan). The final preparations made should be of a more difficult order, to the end that the pupil, by examination of the literature for himself, may make some approach to realizing the conditions of original research. During this year reading and study are again indispensable. Seminar work in which reports on recent discoveries are presented, and topics of vital interest in the point of view of general chemistry are discussed, will serve for reviewing and deepening the knowledge of the subject. There is far too much so-called instruction in chemistry in our higher institutions

---

lems from the standpoint of semi-original investigation with the rigid criteria applied in real research.

Work having these characteristics serves to clinch the impression made when the corresponding topics were discussed in the introductory course. On the other hand, the common kind of quantitative analysis, which devotes itself exclusively to technique, and can be fairly defined by the number of determinations it includes, is of little value at any stage. It may give some mechanical skill, but it will teach no chemistry.

[1] F. H. Thorp, *Inorganic Chemical Preparations* (Ginn & Co.), and Erdmann-Dunlap, *Introduction to Chemical Preparation* (John Wiley & Sons), are similar works.

which consists solely in technical guidance of experimental work, and neglects entirely the development of the scientific knowledge of the pupils. Experimental work without reading, and exercises which call for no thought, are as useless as food without the intervention of the digestive fluids.

It need not be added that during this time physics and mineralogy, at least, and if possible other sciences, should be pursued, not only on account of their indispensability as sources of illustration in the teaching of chemistry, but also because the future teacher may have to give instruction in some of them. The other studies should include a sufficient amount of German to give a reading knowledge, since it is difficult to pursue the study of chemistry without reference to articles and books in this language.

It seems to me that three years, properly spent, will furnish a knowledge of the science which, considering the demands of the secondary school, will be approximately equivalent to that expected in other subjects. The time, however, must be spent as largely as possible in acquiring, adding to, and throwing side-lights upon general chemistry. Long courses in analysis, while they must be included, at some stage, in the training of technical chemists and investigators, are a misapplication of precious time so far as our purpose is concerned. This adequate training cannot be obtained quickly or without expense. It will require almost continuous work throughout the college course, or an equivalent of this.

The question is, where can the teacher in training secure the needful instruction. Not of a surety in the departments of chemistry of our colleges and normal schools as at present conducted. The chemical curricula of our higher institutions, largely through the influence of tradition, are so filled with a mass of specialized work in stereotyped grooves that proper instruction for teachers is difficult to obtain.[1] Their arrangements seem to be made for

The Present Condition of Higher Instruction in Chemistry.

[1] For a highly interesting discussion of this subject, see Professor Armstrong's Presidential Address before the Chemical Society of Lon-

the purpose of training chemists for agricultural stations or commercial work. The conventional order of general chemistry, followed by qualitative and then quantitative analyses, is unfortunate. The two latter subjects, as ordinarily taught (*cf.* pp. 173, 210), contribute practically nothing to the student's knowledge of the science of chemistry in the broader view. They are almost always, for the most part, purely technical applications of a single aspect of the subject, and during their study so much general chemistry is forgotten that the student really acquires a narrower view of the subject in some respects than he had at the end of the first year. The analyst turned out by this training can do the routine work of a factory. His standing is the same as that of a bookkeeper, and his work requires no more extensive training. His preparation does not fit him to assist in advancing chemical industry, any more than that of the mere bookkeeper fits him to manage an extensive business successfully. The student who intends to become a teacher of chemistry has to pick up the nourishment for the growth of a broad knowledge of the subject from what must be admitted to be a rather sparse vegetation in this point of view, and it is at present only the exceptional student who gets it. I should certainly be at a loss to mention any institution in which an ideal course for teachers is given. Yet, as Professor Nichols says, in his admirable paper on *The Training of Science Teachers*: "No institution, whether it calls itself normal school, college, or university, that does not offer the student opportunities of the kinds just indicated [Nichols' course in physics was on the same lines as that outlined above for chemistry], is fitted for the training of the modern science teacher. No institution, the members of the faculty of which are not *bona fide* men of science,

---

don. JOURNAL OF THE SOCIETY, XLV. (1894), 361; reprinted in NATURE, L. 211. See also Professor John H. Long's address on the Teaching of Chemistry in the Medical Schools of the United States. SCIENCE [N. S.], XIV. 360. Mr. Lachman, in an address on the Improvement of Instruction in Technical Chemistry, utters some very suggestive criticisms of the present methods and sketches a substitute JOUR. SOC. CHEM. INDUSTRY, XX. (1901), 546.

devoting themselves quite as seriously and continuously to research as to routine teaching, can hope to produce in its students those qualities and habits of thought that . . . are essential to the highest type of teacher."

II. The Development of the Teacher during Professional Life.

The teacher cannot afford to settle down and dole out his instruction from the slowly petrifying deposit with which his college provided him. He must follow the new developments of the subject, and continually change his mode of presenting every part of it, in order that it may harmonize with the best thought of chemists. Not only this, however, but he must continually increase his own attainments. The best preparation always seems to have been wonderfully meagre compared to the mass of knowledge which we, as teachers, find indispensable in our work.

What to Read.

The reading of the latest text-book is useful, but the most productive method of study is to take up some topic of interest and pursue it to its limits. A subject like nitric acid and the oxides of nitrogen, for example, when studied first in all the general works, then in the larger books of reference, and finally in the original literature, will be found exceedingly interesting. The study of the various determinations of the ratio of hydrogen to oxygen in water, in spite of the somewhat dry aspect which the mere statement of the subject presents, will be found truly fascinating. The determinations of the atomic weights of aluminium, zinc, and other elements, are highly instructive on account of the precautions employed in their execution. These are mentioned as examples, and the catalogue might be prolonged almost indefinitely without going outside the list of subjects upon which many important papers have been published in the English language.[1]

---

[1] References to books treating fully or with especial clearness of many chemical questions will be found scattered through the present work. A large number are given in Newell's *Teachers' Supplement*. The following are a very few references to important and interesting original articles in English.

Action of metals on nitric acid. Freer, *Inorganic Chemistry*, chap-

Not only is the reading of original papers easy after such preparation as the examination of the text-books gives, but, contrary to the popular impression, it is vastly more interesting and incomparably more valuable than the study of books alone. If we want to know about a plant, we must consider the whole structure, and the whole course of development from the seed to maturity. The structure of a few dead chips is as little enticing or useful in this connection, as the study of text-books is in giving a genuine knowledge of what constitutes the science of chemistry. Only the examination of the literature can show us the growth of each fragment of the science and how additions to human knowledge of permanent value are really made. The atmosphere of the text-book suggests the museum or the tomb to one who has breathed the air of the workshop and of life in the original reports of the investigator.

---

ter XXVI. AM. CHEM. JOUR., XV. 71; XVII. 18; XVIII. 587; XXI. 377.

Atomic weight of oxygen. Cooke and Richards, AM. CHEM. JOUR., X. 81 and 191. Keiser, *ibid.*, X. 249; XX. 733. Noyes, *ibid.*, XII. 441.

Atomic weight of zinc. Morse, AM. CHEM. JOUR., X. 311. Clarke, *ibid.*, III. 263.

Molecular weight of hydrogen fluoride. Mallett, AM. CHEM. JOUR., III. 189.

Persulphates. Marshall, JOUR. CHEM. SOC., LIX. 772. JOUR. SOC. CHEM. INDUSTRY, XVI. No. 5.

Nickel carbonyl. Mond, Langer, and Quincke, JOUR. CHEM. SOC., LVII. 750.

Allotropic forms of silver. Carey Lea, AM. JOUR. OF SCI., [3], XXXVII. 476. Barus, *ibid.*, XLVIII. 451.

Absence of chemical action in absence of water. Baker, JOUR. CHEM. SOC., LXV. 611. Shenstone, *ibid.*, LXXI. 471.

Argon. Rayleigh and Ramsay, AM. CHEM. JOUR., XVII. 225.

Helium. Ramsay, JOUR. CHEM. SOC., LXVII. 684 and 1107.

Urea and ammonium cyanate. Walker, JOUR. CHEM. SOC., LXVII. 746; LXIX. 193; LXXI. 489; LXXVII. 21.

Perchloric anhydride. Michael and Conn, AM. CHEM. JOUR., XXIII. 444.

Adsorption. Walker, JOUR. CHEM. SOC., LXIX. 1334.

Flame. Smithells, JOUR. CHEM. SOC., LXI. (1892), 204; LXVII. (1895), 1049; NATURE, XLIX. (1893), 86, also correspondence on pp. 100, 149, 171, 172, 198; CHEMICAL NEWS, LXVI. (1893), 139, 160. Lewis, CHEMICAL NEWS, LXV. (1892), 112, 125; LXVI. (1893), 99.

Reading, however, is not sufficient; there should be continual experimental work adapted to the previous training of the teacher. Making recently discovered compounds, and repeating new ways of making old ones, will furnish opportunities for work of any degree of ease or difficulty. The persulphates, nickel carbonyl, the allotropic forms of metallic silver, and many other interesting bodies can be made with the resources of any laboratory. Some of Baker's experiments on the absence of chemical union in dry materials will give opportunity for the use of experimental skill. If the teacher lacks preparation for this kind of work, he may add to his knowledge by a systematic course of experiments and reading in inorganic preparations, in organic chemistry or in some of the experimentally simpler parts of physical chemistry, such as the observation of the boiling point and freezing point of solutions, the measurement of vapour densities, etc.[1]

Experimental Work.

For the teacher who has the necessary qualifications, the very best exercise of his powers will be in making simple original investigations. I should hesitate to mention this if it were not that Professor Nichols[2] insists upon it as an indispensable feature in the life of every teacher, and that Professor Ganong, in his *Teaching Botanist* (48), makes a strong plea of the same kind. It is well known that the time of the teacher is very fully occupied, and that his equipment is often far from adequate, even for the needs of elementary instruction. It must be admitted, however, that, as Professor Nichols explains at great length, these objections are not conclusive. The professor in the college or university, when we

Research.

---

[1] Much highly instructive work of a kind a little above the ordinary laboratory course in general chemistry is described in Muir and Carnegie's *Practical Chemistry* (Cambridge University Press, 1887), particularly in Part I., Chapters XVI., XVIII.; Part II., Chapters IV.–VII.; Part III., Chapters II.–IV. Most phases of physical chemistry, with the exception of the theory of solutions, are illustrated in these chapters.

[2] *Loc. cit.* See also, for subjects of research in physics, SCHOOL SCIENCE, I. 10.

consider the burden of laboratory teaching and of executive work which he must carry, is on the average no better off than the teacher in the secondary school, and he is expected to pursue research continuously. Nor is elaborate outfit or apparatus necessarily required. There are problems, possibly of a minor nature, which can be solved with nothing beyond the material used in teaching elementary chemistry, unless it be a balance. Even this, however, is not always indispensable. Above all, we must remember that some of the best scientific work has been done, in secondary schools as well as in colleges, by men who had neither time nor appliances which would have encouraged us to expect any productive work whatever.[1]

Summer Schools.

The other means which are available for assistance in the development of the teacher may be mentioned more briefly. It is not often possible for him to take graduate work in some university, but the summer schools, which are now so numerous as to be readily accessible to every one, are taken advantage of by teachers, in some instances, to so remarkable an extent that their power to aid them cannot be doubted. Even if little knowledge, measured by some standards, can be acquired in six or eight weeks, the stimulus and inspiration received by contact with some master of the subject may, even in a brief time, bring forth new life in the teacher who was dying from isolation, and give new vitality to his whole thought and work.

The word isolation reminds us that no efforts of a single individual can ward off for a long time the inevitable petrifaction. Contact with other people with like interests is indispensable. For this reason the meetings of local scientific and educational societies, and the conventions of the American Association for the Advancement of Science and the American Chemical

---

[1] For the encouragement of this work amongst its teachers, the Board of Education of Chicago pays the expenses of any investigations they may make, provided the results on whose accomplishment the claims are based are certified by some competent authority to be genuine additions to knowledge. This most enlightened policy might be with advantage imitated by other school authorities in the country.

Society furnish opportunities of receiving help which should not be missed. Visiting other schools and watching the work of other teachers should also be indulged in as frequently as possible.

No Preparation too Great for the Task.

As has been said before, the teaching of beginning chemistry is the most difficult task which the chemist, no matter what his training, can undertake. Teaching it in a secondary school is more difficult than teaching it in a university, and incomparably more difficult than giving instruction in some advanced branch of the subject, or, assuming proportionate preparation of the teacher, even supervising the work of students engaged in research. These tasks are all different and require perhaps somewhat different qualifications, but the delicate operation of dealing with a young pupil who is beginning the study of a science, so as to impart to the small change of the subject the ring of the genuine metal and the stamp of truth and authority, requires a breadth and at the same time a minuteness of knowledge which only long training and experience can give. The maturity and resourcefulness which are born of a thorough control of the subject cannot be communicated. They are the fruit of unremitting and long continued labour.

III. Literature for the Teacher.

It is impossible here to mention all even of the important works dealing with every branch of chemistry. In the following bibliography the titles have been selected in the main with reference to the needs of the teacher. There are included, however, a number which are adapted also to the use of pupils. A few volumes should be added yearly to the reference shelf in the laboratory, in order that encouragement to excursions outside the narrow limits of the regular text-book may not be wanting. The books have been classified and, under each head, after some remarks in regard to sources of information on the particular branch of the subject, the bibliographical description of commendable works is given. In the case of many topics the appropriate references have appeared already in earlier chapters.[1]

Dictionaries, etc: —Watts' *Dictionary* contains articles varying in length from a few lines to many pages on every chemical substance and every topic in scientific chemistry. Technological subjects have been relegated to Thorpe's *Dictionary.* Extensive tables including much indispensable information will be found in the *Chemiker Kalendar* and Meade's *Pocket Manual.* The articles in the *Encyclopedia Britannica* and other works of the same class frequently treat subjects hardly noticed in text-books. In using them due regard must be paid to the time at which the articles were written.

Watts. Dictionary of Chemistry. Edited by Morley and Muir. 4 vols., half leather. London and New York, Longmans, Green & Co. 1894.

Thorpe, T. E. Dictionary of Applied Chemistry. 3 vols., half leather. London and New York, Longmans, Green & Co. 1894–95.

Biedermann. Chemiker Kalendar. Berlin, Springer. Annually.

Meade. The Chemists' Pocket Manual. Easton, Pa., Chemical Pub. Co. 1900.

Inorganic Chemistry, Larger Works: —The most useful, extensive work of reference is the inorganic portion of Roscoe and Schorlemmer's *Treatise.* It has recently been brought up to date. The other works, which may be classed as university text-books, have each well-defined merits of their own. Remsen is notable for lucidity; Newth, for attention to industries; Freer, for the treatment of certain chapters; Richter, for the remarkable amount of information it gives for its size; Ramsay, for the arrangement of the material. Ostwald's *Outlines* is an attempt to apply the latest developments of physical chemistry to inorganic chemistry, and is highly suggestive. The small

[1] The reader is referred for references on the following topics to the appropriate parts of this book: Elementary text-books on inorganic chemistry, pp. 55–60; Laboratory manuals, pp. 104, 113, 115–119, 192, 216; Questions and problems, pp. 133, 136; Inorganic preparations, p. 211; Lecture experiments, pp. 134, 167, 169, 170; Glassworking and technique, p. 113. Fundamental conceptions of the scientific method, pp. 147–153.

Modern Chemistry of Ramsay is a highly successful attempt to give a bird's-eye view of the same aspect of the subject. The teacher should have as many of these books as possible at his command.

Roscoe and Schorlemmer. Treatise on Chemistry. Vols. I. and II., Inorganic. London, Macmillan. New York, D. Appleton & Co. 1898.

Mendeleeff-Greenaway. Principles of Chemistry. 2 vols. London and New York, Longmans, Green & Co. 1897.

Remsen. Chemistry, Advanced Course. New York, Henry Holt & Co. London, Macmillan. 1898.

Newth. Text-Book of Inorganic Chemistry. London and New York, Longmans, Green & Co. 1897.

Freer. General Inorganic Chemistry. Boston, Allyn & Bacon. 1894.

Richter-Smith. Inorganic Chemistry. Philadelphia, Blakiston. London, Kegan Paul, Trench & Co. 1900.

Ramsay. A System of Inorganic Chemistry. London, Churchill. 1891.

Bloxam. Inorganic and Organic Chemistry. London, Churchill. Philadelphia, Blakiston. 1901.

Ostwald-Findlay. Principles of Inorganic Chemistry. London and New York, Macmillan. 1902.

Ramsay. Modern Chemistry. Part I., Theoretical; Part II., Systematic. London, J. M. Dent. New York, Macmillan. 1901.

Theoretical: — Walker's *Physical Chemistry* is generally held to give the clearest account of the subject which has so far appeared. Lehfeldt's is less well known. It is wonderfully comprehensive for its size, and well balanced in the relative space given to different topics. Dobbin and Walker is elementary. The teacher is advised to study several works on this subject, including some on special parts of the subject like those appended to the list, as it is in this way only that a clear understanding of the theory can be obtained. The second and third last books are reprints of original papers, and the last contains a description of some laboratory methods.

Walker. Introduction to Physical Chemistry. London and New York, Macmillan. 1899.

Lehfeldt. Text-Book of Physical Chemistry. London, Edward Arnold. New York, Longmans, Green & Co. 1899.

Nernst-Palmer. Theoretical Chemistry. New York and London, Macmillan. 1895.

Dobbin and Walker. Chemical Theory for Beginners. London and New York, Macmillan. 1892.

Morgan. Elements of Physical Chemistry. New York, John Wiley & Sons. 1899.

Jones. The Elements of Physical Chemistry. New York and London, Macmillan. 1902.

Jones. The Theory of Electrolytic Dissociation. New York and London, Macmillan. 1900.

Ostwald-Muir. Solutions. London and New York, Longmans, Green & Co. 1891.

Le Blanc-Whitney. Elements of Electro-Chemistry. New York and London, Macmillan. 1896.

Lüpke-Muir. Elements of Electro-Chemistry. London, Grevel & Co. Philadelphia, Lippincott. 1897.

Pfeffer-Van 't Hoff-Arrhenius-Raoult-Jones. The Modern Theory of Solution. New York, American Book Co. 1899.

Faraday-Hittorf-Kohlrausch-Goodwin. Fundamental Laws of Electrolytic Conduction. New York, American Book Co. 1899.

Jones. The Freezing Point, Boiling Point, and Conductivity Methods. Easton, Pa., Chemical Pub. Co. 1897.

Of an entirely different character are the three following books. They do not profess to give much or, in the cases of the two last, any attention to the theory of solutions. They discuss the atomic theory, the constitution of chemical substances, the periodic law, and other subjects, with a strong infusion of the historical method in their mode of treating them. They will be found exceedingly valuable.

Tilden. Introduction to the Study of Chemical Philosophy. London and New York, Longmans, Green & Co. 1902.

Remsen. Principles of Theoretical Chemistry. Philadelphia, Lea Bros. & Co. London, Bailliere, Tindall & Cox. 1892.

Lothar Meyer. Outlines of Theoretical Chemistry. London and New York, Longmans, Green & Co. 1899.

Historical: — The works named below divide themselves into four sets: the general treatises on the history of the science, histories of special periods or special parts of the science, biographical works, and reprints of memoirs of historical interest. Of the books in the second set, Ramsay's *Gases of the Atmosphere* is a useful supplement to the treatment of the air, and particularly of oxygen, as it is found in the text-books. Carnegie treats some selected topics in a very suggestive manner.

The Alembic Club Reprints, the last set, supply some papers of historical interest in a neat and inexpensive form. Study of these documents gives a vivid impression of the attitude and methods of the early workers which cannot be obtained excepting by reading their own descriptions of their labours.

Tilden. A Short History of the Progress of Scientific Chemistry in Our Own Times. London and New York, Longmans, Green & Co. 1899.

von Meyer-McGowan. History of Chemistry. London and New York, Macmillan. 1891.

Ladenburg-Dobbin. Lectures on the History of the Development of Chemistry Since the Time of Lavoisier. Edinburgh, The Alembic Club, Wm. F. Clay (Agent). 1900.

Venable. A Short History of Chemistry. Boston, D. C. Heath & Co. 1894.

Muir. The Alchemical Essence and the Chemical Element. London and New York, Longmans, Green & Co. 1894.

Rodwell. The Birth of Chemistry. London and New York, Macmillan. 1874.

Thorpe. Chemistry in Britain in the XIX. Century. London, JOURNAL OF THE CHEMICAL SOCIETY, LXXVII. (1900), 562.

Ramsay. The Gases of the Atmosphere. London and New York, Macmillan. 1896.

Carnegie. Law and Theory in Chemistry. London and New York, Longmans, Green & Co. 1894.

Wurtz. The Atomic Theory. London, Kegan Paul, Trench & Co. New York, D. Appleton & Co. 1891.

Venable. The Development of the Periodic Law. Easton, Pa., Chemical Pub. Co. 1898.

Thorpe. Essays in Historical Chemistry. London and New York, Macmillan. 1894.

Tyndall. Faraday as a Discoverer. London, Longmans, Green & Co. New York, D. Appleton & Co. 1894.

Muir. Heroes of Science, — Chemists. London, S. P. C. K. New York, E. and J. B. Young & Co. 1883.

Thorpe. Humphrey Davy. Century Science Series. London and New York, Macmillan. 1896.

Roscoe. John Dalton. Century Science Series. London and New York, Macmillan. 1895.

Thompson. Michael Faraday. Century Science Series. London and New York, Macmillan. 1899.

Shenstone. Justus von Liebig. Century Science Series. London and New York, Macmillan. 1895.

Mallett. Memorial Lecture on Stas. London, JOUR. CHEM. SOC., LXIII. (1893), 1; JOUR. AM. CHEM. SOC., Sept. 1892.

Playfair-Abel-Perkins-Armstrong. Memorial Addresses on Hofmann. London, JOUR. CHEM. SOC., LXIX. (1896), 575-732.

Japp. Memorial Lecture on Kekule. London, JOUR. CHEM. SOC., LXXIII. (1898), 97.

Roscoe. Memorial Lecture on Bunsen. London, JOUR. CHEM. SOC., LXXVII. (1900), 513.

Chemical Society of London. Twelve Memorial Addresses (collected). London, Gurney & Jackson. 1901.

Alembic Club Reprints. London, Simpkin, Marshall and Co.; Chicago, The University of Chicago Press.

1. **Black.** Experiments upon Magnesia Alba, etc.
2. **Dalton, Wollaston, and Thomson.** Foundations of the Atomic Theory.
3. **Cavendish.** Experiments on Air.
4. **Dalton, Gay-Lussac, and Avogadro.** Foundations of the Molecular Theory.
5. **Hooke.** Extracts from Micrographia.
6. **Davy.** The Decomposition of the Alkalies and Alkaline Earths.
7. **Priestley.** The Discovery of Oxygen.
8. **Scheele.** The Discovery of Oxygen.
9. **Davy.** The Elementary Nature of Chlorine.
10. **Graham.** Researches on the Arseniates, Phosphates, and Modifications of Phosphoric Acid.
11. **Jean Rey.** On an Enquiry into the Cause Wherefore Tin and Lead Increase in Weight on Calcination.
12. **Faraday.** The Liquefaction of Gases.
13. **Scheele. Berthollet, Morveau, Gay-Lussac, and Thenard.** The Early History of Chlorine.
14. **Pasteur.** Researches on the Molecular Asymmetry of Natural Organic Products.
15. **Kolbe.** Papers on the Electrolysis of Organic Compounds.

Reprints of Science Classics. Chicago: The School Science Press.

No. 1. **Lavoisier.** The Analysis of Air and Water. Tr. by C. E. Linebarger. 1902.

Organic:—The chemistry of the carbon compounds is treated most comprehensively in the new edition of Richter's *Organic Chemistry*. Remsen's work gives an elementary account of the subject and describes illustrative experiments. Hjelt gives a survey of the generalizations of organic chemistry.

Richter-Smith. Organic Chemistry. 2 vols. Philadelphia, Blakiston London, Kegan Paul, Trench & Co. 1900.

Remsen. Introduction to the Study of the Compounds of Carbon. Boston, D. C. Heath & Co. London, Macmillan. 1895.

Perkin and Kipping. Organic Chemistry. 2 vols. Edinburgh, Chambers. Philadelphia, Lippincott. 1894.

Hjelt-Tingle. Principles of General Organic Chemistry. London and New York, Longmans, Green & Co. 1895.

For laboratory work in organic chemistry, the collections of selected preparations by Noyes and by Gattermann are excellent. Of a more elementary character are Garrett and Harden, Orndorff, and Turpin. A compendium of all organic methods of work, with copious illustrations of their application, and numerous references to the original literature, will be found in Lassar-Cohn. Noyes and Mulliken's book gives a different and highly instructive view of the subject.

Noyes. Organic Chemistry for the Laboratory. Easton, Pa., Chemical Pub. Co. 1897.

Gattermann-Shober. Practical Methods of Organic Chemistry. London and New York, Macmillan. 1901.

Garrett and Harden. Elementary Course of Practical Organic Chemistry. London and New York, Longmans, Green & Co. 1897.

Orndorff. Laboratory Manual of Organic Chemistry. Boston, D. C. Heath & Co. 1893.

Lassar-Cohn-Smith. Laboratory Manual of Organic Chemistry. London and New York, Macmillan. 1895.

Noyes and Mulliken. Laboratory Experiments on the Class-Reactions and Identification of Organic Substances. Easton, Pa., Chemical Pub. Co.

Industrial: — Thorp's is the most recent work on the subject. It includes all industries excepting the metallurgical. Borchers' work gives an excellent account of the recent applications of electricity in technological chemistry.

Thorp, F. H. Outlines of Industrial Chemistry. London and New York, Macmillan. 1899.

Huntington and McMillan. Metals. London and New York, Longmans, Green & Co. 1897.

Borchers. Electro-Smelting and Refining. London, C. Griffin & Co. Philadelphia, Lippincott. 1897.

Wagner. Manual of Chemical Technology. London, Churchill. New York, D. Appleton & Co. 1895.

Analytical: — The standard works of reference on this subject are those of Fresenius. The most satisfactory treatment of both branches in one volume is represented by Newth's book. Perkin's work gives special attention to organic analysis. Ostwald's *Scientific Foundation* is indispensable, whatever other works are employed, as none of the treatises on analysis pay sufficient attention to the theory, and most pay no attention to it whatever.

Ostwald-McGowan. Scientific Foundations of Analytical Chemistry. London and New York, Macmillan. 1900.

Fresenius. Manual of Qualitative Analysis. London, Churchill. New York, John Wiley & Sons. 1890.

Fresenius. Quantitative Chemical Analysis. London, Churchill. New York, John Wiley & Sons. 1881.

Newth. Chemical Analysis, Qualitative and Quantitative. London and New York, Longmans, Green & Co. 1898.

Noyes, W. A. Elements of Qualitative Analysis. New York, Henry Holt & Co. 1901.

Perkin, F. M. Qualitative Chemical Analysis. London and New York, Longmans, Green & Co. 1901.

Noyes, A. A. Qualitative Chemical Analysis. London and New York, Macmillan. 1899.

Clowes and Coleman. Elementary Quantitative Chemical Analysis. London (4th ed. 1897), Churchill. Philadelphia, Blakiston.

Sutton. Handbook of Volumetric Analysis. London, Churchill. Philadelphia, Blakiston. 1890.

Thornton and Pearson. Notes on Volumetric Analysis. London and New York, Longmans, Green & Co. 1898.

Hempel-Dennis. Elements of Gas Analysis. London and New York. Macmillan. 1891.

Mason. Examination of Water. New York, John Wiley & Sons. London, Chapman & Hall. 1899.

Blair. The Chemical Analysis of Iron. Philadelphia, Lippincott. 1902.

Smith, E. F. Electro-Chemical Analysis. Philadelphia, Blakiston. 1894.

Classen. Quantitative Chemical Analysis by Electrolysis. New York, John Wiley & Sons. London, Chapman & Hall. 1898.

Landauer-Tingle. Spectrum Analysis. New York, John Wiley & Sons. London, Chapman & Hall. 1898.

Chemistry of Daily Life:—Information about the chemistry of common things is scattered through an immense range of literature. Works on special branches of analysis and on special industries, works on botany, physiology, etc., and many others can contribute much to a knowledge of this. The following profess to deal with such matters in a popular way.

Johnston. Chemistry of Common Life. London, Blackwood. New York, D. Appleton & Co. 1879.

Lassar-Cohn-Muir. Chemistry of Daily Life. London, Grevell & Co. Philadelphia, Lippincott. 1898.

Martin. Story of a Piece of Coal. London, Geo. Newnes. New York, D. Appleton & Co. 1896.

Faraday. Chemical History of a Candle. London, Chatto & Windus. New York, Harper & Brothers. 1862.

Williams. The Chemistry of Cooking. London, Chatto & Windus. New York, Appleton & Co. 1885.

Richards and Elliott. Chemistry of Cooking and Cleaning. Boston, Home Science Pub. Co. 1897.

Richards. Food Material and their Adulterations. Boston, Home Science Pub. Co. 1886.

King. The Soil. London and New York, Macmillan. 1899.

Roberts. The Fertility of Land. New York and London, Macmillan. 1897.

Miscellaneous : — From the works on the many branches of chemistry which have not been treated separately, a few titles have been selected. The bibliography of the New England Association of Chemistry Teachers, to which I am indebted for some of the data in these lists, gives a brief description of the nature of each of the books contained in it. It will be found very useful.

Williams. Elements of Crystallography. New York, Henry Holt & Co. 1892.

Bauerman. Descriptive Mineralogy. London and New York, Longmans, Green & Co.

Dana, E. S. A Text-Book of Mineralogy. New York, John Wiley & Sons. London, Chapman & Hall. 1898.

Abney. Treatise on Photography. London and New York, Longmans, Green & Co. 1901.

Meldola. Chemistry of Photography. London and New York, Macmillan. 1889.

Halliburton. Essentials of Chemical Physiology. London and New York, Longmans, Green & Co. 1901.

Hueppe-Jordan. Principles of Bacteriology. Chicago, Open Court Pub. Co. London, Kegan Paul, Trench & Co. 1899.

Frankland. Our Secret Friends and Foes. London, S. P. C. K. New York, E. & J. B. Young & Co. 1897.

Schützenberger. On Fermentation. London, Kegan Paul, Trench & Co. New York, D. Appleton & Co. 1889.

New England Association of Chemistry Teachers. List of Books in Chemistry. Boston, L. E. Knott Apparatus Co. 1900.

Periodicals: — The best way to keep in touch with chemical work is to read at least one journal regularly. The first five on the list publish original articles. In addition to this, the second contains reviews of all the chemical research done in America. The third contains reviews of all chemical memoirs, wherever published. The fourth is admirably edited, and furnishes excellent abstracts of a large amount of work, even when it is mainly of scientific interest and has little actual bearing on industry. Numbers six to eight publish articles on all the sciences, including chemistry. The last three frequently contain articles dealing with the teaching of chemistry.

American Chemical Journal. Baltimore, Md., The Johns Hopkins University Press. Monthly.

Journal of the American Chemical Society. Easton, Pa., Chemical Pub. Co. Monthly.

Journal of the Chemical Society. London, Gurney & Jackson. Monthly.

Journal of the Society of Chemical Industry. London, Eyre & Spottiswoode. Monthly.

Chemical News. London, E. J. Davey. Weekly.

Science. New York, Macmillan. Weekly.

Nature. London and New York, Macmillan. Weekly.

Popular Science Monthly. New York, McClure, Phillips & Co.

School Science. Chicago, 2059 E. 72nd Place. Monthly.

School Review. Chicago, The University of Chicago Press. Monthly.

Zeitschrift für den physikalischen und chemischen Unterricht. Berlin.

THE TEACHING OF PHYSICS IN THE SECONDARY SCHOOL

Prefatory Note

In writing the first four chapters on the teaching of physics the author has had in mind especially the school-teacher, from the time when, perhaps only a boy, he is making choice of a profession to the time of his full career in charge of a well appointed school laboratory and class room. The motives and considerations which should influence the choice of this career, the academic and other preparation which the prospective teacher should make for it, the means by which he may keep himself in continual progress as a teacher, and the kind of practical problem in which he, without undertaking what is commonly called original research, may find profitable employment for any amount of energy in the improvement of his work, are all touched upon in these four chapters.

In the next chapter the change of aim and method in school physics teaching during the past twenty years is briefly discussed in connection with changes in text-books. This leads naturally to a consideration, in Chapter VI., of the proper general spirit and method of laboratory instruction in schools. The next two chapters deal, respectively, with technicalities of laboratory management, and with the very important functions of lectures and recitations in connection with laboratory work.

In Chapter IX. the possibilities of physics teaching in primary and grammar schools are taken up. In the next chapter attention is given to physics in secondary schools, and the question is raised whether, after all, in view of their probable difference in scholarly quality, the boy who is going to college and the boy who is not going to college should follow the same course of physics in school, or, rather, whether the distinctively preparatory school on the one hand and the high school on the other hand should have just the same kind of physics teaching and work.

Chapter XI., On The Presentation of Dynamics, is the only chapter in the book which is devoted to any one part, exclusively, of physics, the exception in this case being justified, in the opinion of the author, by the exceptional difficulty and importance of the subject of dynamics.

Chapter XII. gives a plan of rooms and fittings for a school department of physics, and Chapter XIII., the last, gives some account of the state of physics teaching in the schools of Germany, England, and France.

The book assumes throughout that the system of physics instruction by combined laboratory and class room work is now permanently established for the better class of American secondary schools; and the author believes it to be the especial privilege and duty of American teachers of physics so to develop and perfect this system as to make it not only a great benefit and advantage to ourselves, but a model for imitation by the schools of Europe, most of which, on the Continent at least, have hardly ventured as yet upon the experiment which we are here working out to a successful conclusion.

In the bibliography which is distributed among these chapters the author has certainly not included all the good books, and he does not feel sure that he has left out all the bad ones. Comments on the various text-books named are given in very few cases, the fact being that, according to the author's experience, no one knows thoroughly the possibilities of a book for good or evil till he has taken a class through it.

Writing these chapters has interested the author and has improved his own teaching. He hopes that reading them may be equally beneficial to others.

EDWIN H. HALL.

CAMBRIDGE, MASS.
March, 1902.

The

Teaching of Physics in the Secondary School

CHAPTER I

WHETHER TO BE A TEACHER OF PHYSICS

REFERENCES.

Eliot, C. W. What Is A Liberal Education? The Century, June, 1884. Educational Reform. New York, The Century Co. 1898.

Fitch, Sir Joshua. Thomas and Matthew Arnold, Great Educators Series. London, W. Heinemann, Charles Scribner's Sons. 1897. Pp. 277.

Hart, A. B. The Teacher as a Professional Expert. SCHOOL REVIEW, I. 4–14.

Huxley, T. H. Science and Education, Vol. III. of the Essays. London, Macmillan & Co. New York, D. Appleton & Co. 1894.

Spencer, Herbert. Education. London, Williams & Norgate. New York, D. Appleton & Co.

Welldon, J. E. C. The Teacher's Training of Himself. Contemporary Review. March, 1893. Pp. 369–386.

As in every other department of pedagogic art, there is in physics the teacher who is born and the teacher who is made. The latter, if successful, is the product of infinite labour, of long-suffering patience with himself, of constant courage, of never-dying willingness to learn; but all this is equally true of any man who aspires to excellence in any art for which his native talent is not conspicuous. By all means, let every man find the thing he can do best, and then do it at his best.

Why should a man be a teacher of any kind? First, the negative reasons, which may be somewhat as follows: Purely manual or clerical work is too limited in its mental scope and is paid too little. The practice of medicine would be too painful or too critically responsible. The pulpit requires one to talk when one may have nothing to say. The bar imposes a professional and mercenary contentiousness for which one may lack both taste and talent. Business success is too doubtful, or is obtainable at too great a price.

Why be a Teacher?

The positive reasons. The profession of teaching is safe, it is honourable; it is, it may be, pre-eminently, absolutely, honest. It brings contact with and influence on young minds in a plastic and growing condition. Viewed with regard to society at large, to civilization, to government, instruction is construction. To use the phrase of President Eliot, the profession of teaching gives a man a chance to "build himself into" the great fabric of the most beneficent and enduring human institutions.

What shall I Teach?

And so we have chosen to be teachers. But what shall I teach? Shall it be one of the ancient languages? The children of the American public are not likely to suffer from too much Latin or Greek. We, perhaps more than any people of Europe, need to be told and shown that it is possible to go too fast straight ahead, that the beautiful, the true, the desirable, may be behind us, that the most wholesome, rational, happy living is consistent with, nay, requires, a certain leisureliness of mind which takes occasion to learn how men have lived, and what they have done that is worth remembering and imitating. On these or similar grounds one may amply justify the choice of any great language, or literature, or history, as the sphere of his life-work as a teacher, provided he be not incompetent for the task to which he devotes himself.

Why then should we turn from the "humanities," from the study, in its various phases and achievements, of "this pleasing anxious being," human life, to experimental and mathematical science, for which so few have any further interest than a desire to enjoy its material benefits and to be entertained by its occasioual spectacular displays. Putting aside for the moment the question of individual and special talents, we can see that physical science has for some minds, or some temperaments, a peculiar charm in this, that it holds out to every devotee the possibility of making by himself some positive, absolutely new, addition to the sum total of permanent useful knowledge, the certainty of moving forward into regions of thought and of power which no previous generation of men has ever penetrated since the world began. This motive appeals to the north-pole spirit, of which every true follower of science must have a dash, the spirit which can find pleasure in places where the air is cold but pure, where the footing is rugged but forward. Contrasted with the study or teaching of any language, as such, the study and teaching of science offers, as the object of especial attention, substance instead of form; and though the form in the one case be the expression of human thought, while the substance in the other case is that of things not made by man, yet we who are of the school of science cannot admit, because we do not feel, that we are on lower ground. He only should make such a confession who follows science for its mere utilities.

But, given the intellectual predisposition in favour of science, what special tastes or talents should prompt or justify the choice of physics?

Qualities needed for Physics Teaching.

The most desirable qualities are, in my opinion: First, capacity for clear, sustained, correct thinking, most conveniently tested by capacity for some common branch of mathematics. It is true that Faraday, who was a very great scientific thinker and discoverer, declared, after turning the handle of a calculating machine, that he had now for the first time in his life performed a mathematical operation. It is true that Edison, who is a great scientific man of a different kind, has said that he never could do much with algebra, being bothered by the plus and minus signs. But it is not to be supposed that either of these men was really lacking in mathematical faculty. The fact is that neither of them had, as a boy, much regular education. Each of them was carried by native talent early in life into conspicuous achievements in science, and after that each probably felt, consciously or unconsciously but in either case rightly, that to go back and try to educate himself as others are educated, would be to throw himself off the track of already assured success.

Moreover, even if we admit that mathematical ability is not absolutely essential to success as a teacher or investigator in physics, we must find that the literature of physics, as shown in text-books and in periodicals, is so permeated with the ideas and the symbols of mathematics, that a person who at the outset must confess to a weakness with respect to such ideas and symbols would enter the advanced study of this literature under a heavy handicap.

The question remains whether it is indispensable that the prospective school-teacher of physics shall look forward to what would be called, among physicists, advanced study. This question may be frankly answered in the negative. One can get enough of physics without knowing anything of the calculus to be a good school-teacher of this science. No energetic man who wishes to be such a teacher need be deterred by lack of interest in mathematics, provided he has the endowment, which I put second, with some doubt whether it should not be put first, of capacity for a quick understanding of machinery. This may show itself in achievements ranging from the easy mastery of a mouse-trap to the easy mastery of a compound steam-engine. I dwell upon facility here, for the reason that with facility goes liking, and with liking goes knowledge, and a wide acquaintance with machinery and apparatus is useful, not only in equipping and maintaining a laboratory, but also in awakening the interest and holding the attention of pupils, who, whether mechanically competent or otherwise, always admire mechanical proficiency in others. The man who is slow to think out the relations and working of a machine may in time acquire a competent knowledge of such apparatus and machinery as comes within his range of habitual vision, but he is sure to have an occasional bad five minutes in the presence of his class during the early years of his teaching.

I assume, although this is not always true, that a considerable degree of manual skill and proficiency in the use of tools will accompany the instinct for machinery.

No other qualities than the two now briefly discussed need be mentioned as important for physics especially, though of course all the intellectual virtues count here, as they do in other teaching. A great memory for facts in detail, such as the chemist needs, the habit of minute general observation, so nearly indispensable to the naturalist, — these traits are certainly useful to the physicist, but he can do without them. Inventiveness, constructive imagination, is eminently desirable; but a reasonable measure of it is pretty sure to be associated with the mechanical faculty already spoken of, and an unreasonable measure of it, which we sometimes find, makes its possessor troublesome, because he will not be content to do anything as other people do it, but must invent his own methods for every operation, out of a mere wantonness of originality. Necessity is not the only mother of inventions.

CHAPTER II

PREPARATION FOR TEACHING

REFERENCES.

Barnett, P. A. Common Sense in Education and Teaching. London and New York, Longmans, Green & Co. 1899. Pp. 321.

Cajori, Florian. A History of Physics. London and New York, Macmillan. 1899. Pp. 322.

Lodge, O. J. Pioneers of Science. London and New York, Macmillan. 1893. Pp. 404.

Monroe, Will S. Bibliography of Education, International Educational Series. New York, D. Appleton & Co.

Munroe, James P. The Educational Ideal. Boston, Heath & Co. 1895. Pp. 262.

Osborn. From the Greeks to Darwin. London and New York, Macmillan. 1896. Pp. 259.

Venable, W. H. Let Him First Be a Man. Boston, Lee and Shepard. Pp. 274.

The "Scientific Memoirs" issued by the American Book Company (New York), under the general editorship of Professor J. S. Ames, are historically interesting and valuable. They are reprints of the original papers announcing great discoveries or important general laws, and are accompanied by biographical sketches of the authors.

Knowledge of the Subject.

THE first thing to be considered under this heading is how to get a competent knowledge of the subject-matter and the methods of the science. This cannot be done without much text-book study and much laboratory experience. The best arrangement of work combines these two methods of training, from the school days on to the attainment of the final degree, and even through all of one's professional life ; for the two relieve and supplement each other, and the time never comes when the aspiring teacher can say, I know enough for my work, I will be a student no more, or when he can, to the best advantage, learn by print alone.

But there is no one necessary arrangement and order of preparation in science. The young man whose undergraduate days are spent in a small college, where the opportunities for laboratory work may be very limited, should make the most of his general course in physics, which will probably give him the reading of a good text-book, and the seeing of a greater or less number of instructive lecture-table experiments, should attend faithfully to his mathematics, carrying this study as far as circumstances will permit, and should get what he can of chemistry. If he has done all this, though he may not yet be a specialist in physics, he will be in excellent condition to appreciate and to profit by the opportunities which he should seek later in the graduate school of some university.

Order and Extent of Study.

But how long should the period of formal study last, how far should it carry the prospective teacher before he begins the practice of his profession? How much, for example, of mathematics is necessary? Must one take the calculus? Without unduly magnifying the importance and solemnity of our profession, without imitating the ambitious example of those who introduce the study of psychology into the curriculum of a cooking-school, we are compelled to pause before answering these questions, and frame some brief philosophy of the objects of education and of the functions of the teacher.

The objects of education in science are, on the one hand, to make men capable, self-sustaining, physically comfortable, on the other hand, to increase their capacity and opportunities for intellectual enjoyment. Each of these objects has an ethical aspect; for men who are materially well-to-do, and intellectually happy, can hardly help being good men and good citizens.

Objects of Education in Science.

If all communities were alike, and all youths in each community alike, if all text-book makers wrote with perfect appreciation of and care for the needs of these young people, providing information, training, stimulus, in due proportion and quantity, if all school boards made ample provision for the use of such books, — if all these conditions held, any one who had learned the text-book and the apparatus described by it, could be a successful teacher, and the profession of teaching would be honoured and paid accordingly. But no one of the conditions mentioned does hold. Communities and schools in America range from large to small, from rich to poor, through many fold, from rural to metropolitan, from the racially homogeneous to the polyglot, from those having traditions of scholarship to those having no traditions at all. As to the individual members of any class in school, some have capabilities for the theoretical side of physics, some for the practical side, some for neither. School boards and school principals may be incapable of making a wise choice of text-books, which vary greatly, not simply from good to bad but in method and purpose.

The Teacher must be a Guide.

Accordingly, the competent teacher is not a mere piece of machinery, made, like an elevator, to run with safety and despatch, carrying its load of passengers through a certain fixed distance along fixed lines and then discharging them, without responsibility or care for their future fate and their ultimate destination. He is, or is prepared to be, a guide, an adviser. However narrow his habitual horizon, he must know what lies beyond it. He must ask, What is the best kind of training, the best kind of information, the best kind of stimulus, for this particular class, for these particular individuals, before me? How can I make this year they spend with me count most toward their life-long efficiency and happiness? How can I best develop their capabilities, correct their worst tendencies, influence their careers? Of course every teacher who attempts all this will fail in much of his endeavour; but it is better that he should try, and that he should make in his student days a preparation adequate to the responsibility which will rest upon him. This means that he must know well all that he will be called upon to teach directly, and have a good general knowledge of much more. He may not be, probably will not be, in active teaching, able to keep up his more advanced studies, if they have ever extended far; but he will, if he has done his work faithfully and intelligently, retain at least an enduring reminiscence, a sustaining memory, that will be a source of strength to himself and of inspiration to his pupils.

Ph D. or A.M.

I am not prepared, as some others may be, to advise that every prospective school-teacher of physics should take the degree of Doctor of Philosophy; for this, requiring ordinarily three or four years of study beyond the baccalaureate course, a high degree of specialization, and much labour devoted to research, is a luxury, a superfluity of preparation for his work, which the school-teacher cannot usually afford. The degree of Master of Arts, with the meaning it is now coming to have, as the certificate of one or two years of graduate study, usually devoted to some specialty, but with little or no original research, seems to me the reasonable goal of the school-teacher in preparation at present.

Need of Mathematics.

I have said elsewhere that one can be a good school-teacher of physics who knows no more of the science than one can get without the calculus, but my advice to the teacher who wishes to realize the possibilities of his profession is strongly against this limitation. Without the calculus one can read almost every page of an ordinary general English treatise on physics, such as Barker, Deschanel, Ganot, Hastings and Beach, Watson, etc., all of Faraday's writings, Maxwell's *Heat*, Tait's *Recent Advances*, Lodge's *Modern Views of Electricity*, and a great deal more excellent literature of physics. But if the student would consult the larger general treatises, or follow the progress of research as revealed in such periodicals as the PHILOSOPHICAL MAGAZINE or the ANNALEN DER PHYSIK, he will find himself baffled and mortified, if he has not a good working knowledge of this mathematical method of developing and expressing physical theories. It is true that a good deal of what is thus hidden from the non-mathematical reader he can perfectly well do without. It is also true that one who has a good knowledge of the calculus is likely to find much that is printed very hard and possibly unprofitable reading. But he who has become familiar with the language of the calculus will always have at least the satisfaction of feeling that he has the key to the gate of knowledge, that he can enter the field that lies before him, however great may be the difficulties that would await him there. This sense of freedom, of possibility, is worth much, even though it may be rarely put to the proof.

Chemistry.

In the training of the teacher of physics should be included a respectable amount of chemistry as well as of mathematics, partly because the chemistry is needed in connection with physics, partly because the teacher of physics is likely to be, at first, if not permanently, a teacher of chemistry also.

Summary of Work.

One "course" of study being counted as the equivalent of one quarter of the work of one college year, the Master of Arts, as I have him in mind, well equipped for the school-teaching of physics or mathematics, and tolerably fitted for the teaching of chemistry, will have taken, in addition to the pre-college physics of a good school, about five courses in physics, two or more courses in chemistry, for the character of which the reader is referred to Chapter VIII. of the first part of this volume, three or more courses in mathematics, including solid geometry, plane trigonometry, analytical geometry, and the elements of calculus with applications to mechanics. This makes ten or more courses of science study, the equivalent of about two and one-half solid years of college work, out of the whole time, at least five years and often more, supposed to pass between admission to college and the attainment of the M.A. degree.

More Specific.

The physics may well include one course or somewhat more in a general text-book, like one of those named earlier in this chapter, with accompanying laboratory exercises of an illustrative and not too exacting character, usually quantitative, but not painfully accurate, a course of careful laboratory work, with much text-book study, in heat and light, two such courses in electricity and magnetism, and a half course in thermodynamics. Every one of these courses except the first will naturally require some use of the calculus.

Somewhere in the curriculum the student should learn to take, and to make use of, the indicator diagram of a steam-engine and the characteristic curves of dynamos. Engineering study in general is, in my opinion, more important for the prospective school-teacher of physics than special research in pure science.

Other Work.

A brief course in mechanical drawing, and another in the use of ordinary wood-working and metal-working tools, should be got in somehow, in the summer if need be, unless the student is already tolerably versed in these arts; for every teacher of physics should be qualified in some measure to describe and make apparatus. The habit of making rather careful drawings approximately to scale, in the designing of anything to be constructed, is one which in the end saves time and trouble and expense. A reasonable acquaintance with tools and with the processes of the workshop often enables one to foresee and to avoid, without sacrifice of anything desirable, difficult and expensive manipulations in the plans which at first occur to the designer. How to make things with the least labour, if he must make them, how to get the most for his money, if he has money to spend, are questions which the teacher must ponder well.

A moderate degree of skill in a few of the simpler operations of glass-blowing should be sought by observation and practice at every reasonable opportunity, and in general the student should give attention not merely to the main tasks which are plainly set before him, but to those sources of extraneous information and experience which, if duly cultivated, will yield a profitable return. The habit of general observation, not of everything under the sun, but of what will bear on one's professional career, cannot be formed too early. Fortunate is he to whom this habit is instinctive; but he to whom this special talent is not given need not despair.

Habit of Observation.

The resolute and persistent will, this is the potentiality of all talents.

History of Physics and Physicists.

In addition, the teacher of physics should know something of the history of the science and of the lives of the men who have especially developed it. A class is always interested to hear, for example, a brief account of the long contest, beginning in the time of Newton, which ended in the final establishment of the undulatory theory of light. Pupils like to know just what Galvani was doing with frogs' legs when he made his immortal discovery. If they go far enough in the study of physics, they will be entertained by the British-German controversy over the merits of Mayer's work on the mechanical equivalent of heat. It is well also to explode the myth about James Watt and the tea-kettle, replacing it by the sufficiently interesting true account of his development of the steam-engine.

Study of the Art of Teaching.

I have said nothing thus far concerning the study, by the prospective teacher, of the art of teaching as such. In common, probably, with most college teachers of physics, I hold rather conservative views in regard to such study. But it would ill become one who is writing a book on the art of teaching physics to maintain that this art cannot be profitably studied through books or from the oral discourse of those who have practised it long. My state of mind in regard to this matter is perhaps one which may as well be frankly analyzed here and now. In the first place, I have my full share of the prejudice created against "methods" by the superficial, ill-balanced work of the early normal schools. In the second place, I hold that the student who has been well taught, has necessarily had, along with his conscious instruction in the science of physics, a good deal of possibly unconscious instruction in the art of teaching physics. In the third place, I have some apprehension lest the conscious study of this art will be accompanied by an over-conscious attention to the philosophy and psychology of the art, with the possible result of setting up a more ponderous system of mental machinery than can be used to advantage in the very practical, common-sense business of teaching young people.

The students in any training school or college will perhaps be more exposed to this danger than their teachers will be; but even the teacher, if his main attention is directed not to any science as such, but to the art of teaching his pupils how to teach their pupils this science, if he habitually looks at his chemistry or his physics through a double layer of more or less opaque humanity, even the teacher runs the risk of ceasing to be what he ought to be, a chemist or a physicist with an inclination toward pedagogy, and becoming the less robust individual who may be described as a pedagogist with an inclination toward chemistry or physics. The science teachers in teachers' colleges should have the ability, the means, and the opportunity, for doing some original work, work of research, in their sciences as such, without any regard, for the time being, to the pedagogic aspect of their profession, but with every sense and faculty steeped in the atmosphere of pure inquiry and bent to the prosecution of the simple scientific end proposed.

The art of teaching is now receiving a great deal of intelligent attention, and progress is evidently being made. The old normal schools, instead of being abolished as useless, have been improved and are therefore growing in public esteem. Such a school as the very flourishing Teachers College of Columbia University is an experiment, or, rather, an experiment station, which must be watched with interest by intelligent educators everywhere. The opportunity of beginning one's teaching in a moderate way, under the supervision of an experienced teacher and frank critic, before taking the full and permanent responsibility of conducting a class, is an opportunity to be desired. If in divinity schools there is place and use for courses in "Homiletics and Pastoral Care," it is difficult to see why, in the nature of things, there should not come to be in the formal training of the teacher a place and use for courses dealing with the technicalities of his art as such. I observe, however, that, in the catalogue from which I have taken the heading quoted above, this heading is placed over the very last division, except elocution, of the courses of instruction open to students of divinity. The learning, the science, stands foremost; then comes the art. Thus should it be in the training of the teacher.

CHAPTER III

THE TEACHER AS STUDENT, OBSERVER AND WRITER

REFERENCES.

Nichols, E. L. Research Work for Physics Teachers. SCIENCE, Feb. 8, 1901.

IN spite of what I have said in the preceding chapter concerning the degree of Doctor of Philosophy, I do not wish to discourage those who feel themselves willing and able to make what I have called a superfluity of preparation for the profession of teaching in schools. To some men the labour of scientific research, with all its inevitable drudgery, with all its large possibilities of negative results, is a joy in itself, a passion not to be quenched by the passage of years or by the lack of visible rewards from without. If a man has this passion, by all means let him cherish it and, so far as he can, gratify it. Let him go on, for example, to the attainment of the doctor's degree, and then, if his life-work proves to be teaching in a school, let him preserve there, so far as he can without neglect of his first duties, his ambition and his habit of scholarship.

Original Research?

Will such a man find opportunity and means for original research while engaged in school-teaching? Perhaps so. The late Professor Rowland once asked a former student of his own whether he was doing any such work. The reply was, "No, I have n't the time, and I have n't the money;" whereupon Rowland exclaimed, "Don't need any time, don't need any money, if you have only got the will." But on another occasion, when he was asked what he should do with his students while making his researches, he replied, " I shall neglect them," which was, in a way, true. It was well for the world that Rowland did neglect his routine work, and he was readily forgiven this shortcoming; but the ordinary doctor of philosophy must not expect, because he will not be able to deserve, a like indulgence.

Professor Nichols's Suggestions.

The real obstacle to research by teachers is not so much lack of time and means for laboratory work, as lack of the genius for finding the right thing to do, and lack of time and opportunity and will for keeping up with the literature of research. Successful research involves, first, hitting upon some more or less important problem; second, making reasonably sure that no one else has solved the problem; third, solving it. In a paper read before the Physics Club of New York, in December, 1900, and published in SCIENCE for February 8, 1901, Professor Nichols of Cornell has discussed the possibilities of research by teachers in a very interesting and suggestive way. I cannot help thinking, however, that this paper is likely to be of more use to college teachers than to school-teachers; for very few men who have not had experience and training in research during their student days are likely to cope successfully with the difficulties presented even by the apparently simple problems which Professor Nichols mentions as within the reach and power of teachers in schools.

The advice given by Professor Nichols, that every teacher of physics should habitually read either SCIENCE ABSTRACTS or the BEIBLÄTTER to the ANNALEN DER PHYSIK, as well as some other standard journal of physics, seems to me excellent in spirit and intention, though possibly a little too sweeping. Much that appears in these periodicals is rather discouragingly beyond the reach of the school-teacher of physics as he now exists, or as he is likely to exist for some years to come. A dry summary of highly technical papers, often contradicting each other, usually given without authoritative criticism or comment, can be inspiring only to him who is in the thick of research, and able to give much of his time to reading and investigation.

Any very important discovery or advance in physics is pretty sure to be noticed before long in more popular publications. NATURE, SCIENCE, and some of the engineering journals will be found more genial and, I believe, more profitable reading than the ABSTRACTS or the BEIBLÄTTER by most teachers in schools at present.

It is of doubtful advantage to any man to undertake a thing which is too hard for him. All of us do best by doing well the things that naturally come to us, from circumstances or from the promptings of our own natures. The surest profit of research, the mental and moral exercise and enjoyment of it, may be attained in full measure by him who has no thought of publication, and whose only conscious object is to improve the quality of his teaching. Is there an habitual experiment or an habitual laboratory exercise that goes badly? If so, just there lies the opportunity and the motive for research; and when this research is successfully ended, not only the teacher's pupils but his fellow-teachers also should profit by it. Is there some natural phenomenon, within doors or without, in the sky, the air, the water, the ground, of which the teacher does not have a satisfactory description and theory; just there is the occasion for first-hand observation and reflection, perhaps for original discovery. The habit of really looking at and thinking about the familiar objects of every-day experience is sure to bring usefulness and may bring fame. Some men have this habit by nature, and may even suffer from a too miscellaneous interest in what goes on about them; but others must deliberately cultivate the habit, directing it toward such things as are, from a professional standpoint, important for them.

Work akin to Research.

I shall presently give, at considerable length, examples of the problems, essentially problems of research, which every teacher of elementary physics will find in his laboratory work, and of the way in which I have attacked these particular difficulties. I say "attacked" advisedly, for I do not claim complete victory, or a finished undertaking. The results reached in this endeavour are not of very great importance in themselves; but they will illustrate well enough the short and halting forward steps by which progress in the art of elementary laboratory teaching is made.

Before giving these illustrations, which will make the next chapter, I wish to enumerate, as examples of the things, outside of books, which may profitably engage the attention of the teacher or his pupils, certain objects and phenomena which have interested me during a recent visit to the seashore. These are: The waves which accompany the progress of a steamer in a straight line through a smooth sea, their shape and succession, the surprising length of time after the steamer has passed before they reach the neighbouring shore, the ripple mark they channel in the sand along the line where each advancing wave meets its retiring predecessor in a swirl; the bits of mirage, by which, at times, distant objects just above the water-line are shown double, as by a horizontal aerial mirror, and the distinction between such reflection and that produced by the water-surface; the smooth patches, "slicks," in ruffled water and their probable cause; the wind-ruffled patches in smooth water and the frequent slowness of their progress; the proper angle between boom and keel in sailing, and how much it is affected by friction and leeway; the floating of sand, in grains and patches, inches across, on the surface of calm water; the functions of the squid's mantle; the differences of stroke which persons of different bodily shape must use in swimming; the best position of the body for swimming in rough water; the presence and circulation of fresh water in the sand near the reach of the tides.

Physics out of Doors.

On these and other things which met my eye and held my thought, I probably made no observation or reflection that had not been made a thousand times before, — none that would be worth reciting or printing at any length; but my days were fuller, my enjoyment wider and more rational, my profit greater, than if I had not seen them and thought about them.

All these things of which I have just been discoursing attracted my attention during a vacation. At home I should have seen few of them and perhaps not have reflected on these few. Change of scene, variety of experience, is for most of us necessary, if we would keep our faculties awake. An occasional change of text-books, for the mere sake of refreshment of mind, or, better, the habitual use of a number of text-books, is wise and wholesome. The habit of coming together for professional consultation with other teachers, of visiting their laboratories and classes, is a habit to be resolutely maintained. The man who lives too much within himself, who moves within too narrow limits of experience and thought, soon comes to dread contact with his fellows, to shrink from new ideas as one whose limbs are cramped shrinks from motion. When the teacher finds himself in this condition, he must rouse himself; to yield is for him to become an old man at once, whatever his age as reckoned by the calendar.

Change and Variety.

Of course, he must not give up his professional methods and ideas merely because they are questioned or criticised or denounced or ridiculed. He must rather hold himself ready to defend them, so long as he believes them worthy of defence; but in order to do this he must be willing to hear what is said against them. Few men, I take it, like to be made a target by critics or adversaries; but those who cannot bear it must be content to remain unseen.

Writing and Speaking.

The practice of writing for publication is to be commended for the teacher, provided there is some professional matter in which he believes himself able to interest and instruct his fellow-teachers or the public at large. It is true that his judgment of what is interesting and instructive may not always be confirmed by that of editors, especially of editors who pay for what they accept; and it is doubtless true that no form of rejection sufficiently suave to be quite satisfactory to the author of the rejected manuscript has yet been found; but, nevertheless, writing is good for the writer. It compels him to think clearly where he may have thought vaguely; it keeps his attention on the theme discussed and rouses thoughts he never knew before or, knowing, has forgotten.

Even the rejection of his offering by some hard-headed, if not hard-hearted, editor may be salutary, impelling him to cultivate a livelier habit of thought and a better style of expression. Such excellence of language, in speech and writing, as the teacher may readily be capable of, he should habitually maintain; for a good style is more intelligible, as well as more pleasing, than a bad one. Every English-speaking teacher should be, directly or indirectly, a teacher of English.

CHAPTER IV

PROBLEMS OF LABORATORY PRACTICE

REFERENCES.

SCHOOL SCIENCE. Chicago. Zeitschrift für den physikalischen und chemischen Unterricht, Berlin.

IN this chapter I have brought together accounts of attempts recently made to improve in certain respects the work done in one of my laboratory classes, work similar to that of the physics classes in many secondary schools. I give these accounts as showing the kind of work, original in its way, in which any teacher may employ his energy for the betterment of his art.

On Water-Proofing Wooden Blocks.

In elementary laboratory work it is a common practice to put wooden blocks or rods into water, and for such purposes that any considerable amount of soaking would be decidedly objectionable. A simple, easily applied, and thoroughly effectual method of water-proofing is, I fear, still to be discovered. Paint, in a very thick coat, would doubtless protect the blocks for a time; but painted blocks would be rather unsightly, and, moreover, the paint would come off in the hard usage of elementary laboratory work. Similar objections apply to any form of varnishing that would really keep out the water. Impregnating the wood with paraffin seems to be, on the whole, the best device. There are various ways and degrees of doing this, and I have been in doubt until recently, as many teachers may still be in doubt, as to the relative efficiency of these various practices.

From my experiments in water-proofing cherry-wood blocks, I have come to the conclusion that, for ordinary laboratory purposes, the use of an exhaust pump to draw out the air from a block submerged in very hot paraffin is not necessary, but that it is well to allow the block to remain buried in the paraffin while cooling. A block about 9 cm. long, with the grain, and about 4 cm. by 3 cm. across the grain, treated in this way, absorbed about 2.7 gm. of paraffin, increasing in weight from 58.1 gm. to 60.8 gm. Later, submerged in water, this block gained in weight about 0.8 gm. in 4½ hours and 2.1 gm. in 17½ hours.

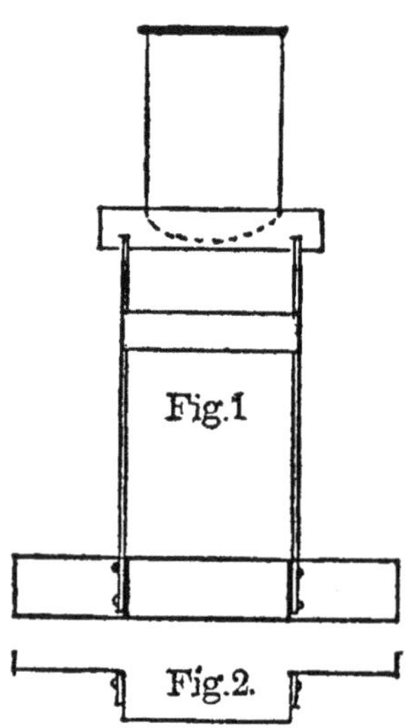

As a result of this study, I have had constructed, for the purpose of waterproofing wooden blocks of a familiar size, about 7.5 cm. square by 3.8 cm. thick, a copper trough 25 cm. long, 9 cm. wide, and about 13 cm. deep, slightly rounded at the bottom so as to allow free movement of paraffin beneath the blocks. Across each end of the trough, and extending a little below the bottom, is soldered a metal bar, the ends of which project on either side. These crossbars serve as a support for the trough when it is placed, for cooling, in water. For heating, the trough is supported by resting the ends of the crossbars in slots at the top of brass posts reaching up from a broad wooden base, the height being so adjusted as to adapt it to the use of a Bunsen burner.

Fig. 1 shows an end view of the trough on its support. Fig. 2 shows one end of wooden base as seen from above. Fig. 3 shows how, by means of a brass strip, *ss*, properly shaped, the blocks may be prevented from floating in the melted paraffin.

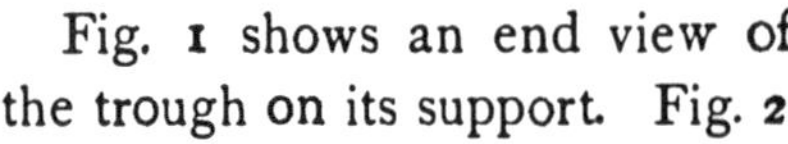

The following seems to be a good method of procedure. Melt enough paraffin in the trough to have a depth, before the blocks are put in, of about 4 cm., and bring this to a temperature of about 100° C. Put in six blocks, each having a stout wire twisted around it to separate it from its neighbours and to serve as a handle for lifting it from the trough; put in the brass strip, *ss*, Fig. 3; heat the bath again to 100° C.; lift the trough from its support and stand it, or float it, in a large tank of cool water; after the trough has been in the water half an hour or more, remove it from the tank, lift out the blocks, which will doubtless be covered in part by solid paraffin, and wipe them carefully while the wax is still soft. In order to take out the blocks with ease, it will probably be necessary to apply the flame for a little while to the bottom, sides, and ends of the trough.

On the Use of the Spring-Balance.

The spring-balance deserves and, I must admit, needs some words of defence and exposition from its friends. It is so inexpensive, so convenient, so quick in its action, that I make very much use of it, in spite of its exasperating inaccuracy in its native state. This inaccuracy is, of course, due to the cheapness of its construction, the price at which it is sold not warranting the adaptation of the scale of each balance to the idiosyncrasies of its spring, so that the faces are merely stamped in divisions corresponding more or less closely to the behaviour of the average spring; and, beyond that measure of adjustment, agreement of spring and scale is a matter of luck.

I have, in the Harvard *Descriptive List*, written out with care directions for studying inaccuracies of the scale and recording them in a graphical form. But further experience and reflection have convinced me that the best thing to do with an incorrect scale is to cover it with paper, fastened on with mucilage after the face of the scale has been cleaned and slightly roughened with rather coarse emery, and mark a new and, as nearly as may be, correct scale on this paper. To do this successfully, well enough at least to improve greatly the graduation of many balances, is an exercise not too long or too difficult for a rather young class. The following method works well: Cover the scale of the balance, which we will suppose to be one of the very familiar 250-gm. instruments marked off in 10-gm. divisions, with paper fastened on with mucilage. Cut a thin strip of spring brass to the shape shown in Fig. 4, the scale of which is ½; bend this strip sharply and carefully along the dotted line *dd*, and so shape it that, when this bend is fitted to the edge of the balance case, the end *ee* will clasp the back of the case with some firmness, as in Fig. 5. The part *dc* should be of such length that *c* will nearly touch the index of

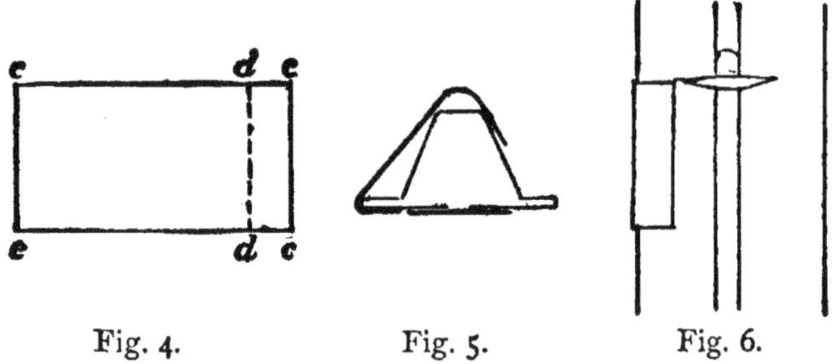

Fig. 4. Fig. 5. Fig. 6.

the balance when the two are on the same level, as in Fig. 6. Now suspend the balance in a free vertical position, without load, and slide the spring-brass clasp, which should grip the case firmly enough to stay in position wherever placed, up or down until one edge is in line with the significant part, point or line, of the index, or, rather, make the adjustment in such a way, that, when a fine pencil line is made along one horizontal edge, this mark will be in line with the significant part of the index. After this zero line is drawn, put on a 10-gm. load, make a new adjustment of the slider, draw the corresponding line, etc. A graduation made in this way ought not to be wrong at any mark more than 1-gm. Change of performance of the spring, from rust or other causes, may in time invalidate this graduation, but it will always be easy to make a new one.

It must be granted that, even after the scale has been corrected, the pupil can easily make errors in the use of the spring-balance, and that, in fact, he must look with some care in order not to make errors of a gram or more in his readings. I do not, however, consider this an objection to this form of balance. The habit of intelligent care is one of the most important objects to be attained in the study of physics or in any other study. A boy who is unwilling or unable to read such a spring balance as I have described, on which the 10-gm. divisions are about 0.2 cm. long, without commonly making an error exceeding 1 gram, is not fit to use a beam-balance; for the latter, though far more sensitive than the spring-balance, is no automaton miraculously contrived to prevent the natural consequences of stupidity and carelessness. In order to yield good results in the hands of some boys, a piece of apparatus must have such cunning that it will not permit itself to be set wrong or read wrong, and such vigour of constitution that it will not mind being knocked off the table occasionally.

In studying the parallelogram of forces, some teachers use weights suspended by cords passing over pulleys, the latter being attached, but movable, along the edge of a round table. The friction of the pulleys and the need of a table of peculiar form seem to me objections to this device. Spring-balances, correctly graduated, are, I believe, preferable in this experiment.

On the Use of the Platform-Balance.

Double-platform balances, costing, with weights, about $7.00, weighing anything up to a kilogram, with an ostensible accuracy of 0.1 gm., are very common and extremely useful pieces of laboratory equipment. Unfortunately, they do not always live up to their professions; and I never feel quite sure of the last tenth of a gram, even if the set of weights accompanying the balance is correct within that limit, which is usually more than doubtful. There are some experiments, which I like to give, in which a change of 0.1 gm. in the weight of some heavy body is rather important; and yet, for use in such experiments, I know nothing which, all things considered, is likely to replace this balance, with all its faults.

As the absolute weight of the body dealt with, in the experiments just mentioned, is usually a matter of small importance, provided its change of weight during an experiment can be correctly found, inaccuracies of the metal weights used are not likely to be troublesome, provided the large weights, in which the perceptible errors are likely to exist, remain in use during the whole operation. The most serious causes of error, and the means for avoiding them, are perhaps sufficiently indicated in the following set of rules for the use of the kind of balance now under discussion:

Directions for the Use of the Double-Platform Balance.

1.[1] When it is necessary to weigh an object to a fraction of a gram on this balance, a vertical scale should be placed on the table alongside the outer edge of one platform, and the balance should be regarded as in equilibrium only when the edge of the platform is level with some particular mark on this scale.

2. Before making any weighing, be sure that the balance is in equilibrium, with no load on either platform, and with the sliding weight at zero. This test should be repeated frequently during the use of the balance.

3. Put the object to be weighed at the middle of the left-hand platform. Put the weights, especially the large ones, as nearly as may be at the middle of the right-hand platform.

4. Any heavy shock suffered by the balance is likely to put the bearings temporarily out of order, enough to affect the weighings perceptibly. Such a shock is most often given by unloading one platform while the other remains loaded. When

[1] Section 1 of these directions has reference especially to balances of the old pattern without pointer. Many balances now have pointers, but rather unsatisfactory ones. If the ordinary pointer were doubled in length, and brought to a sharp tip close to the scale, which should be reduced to a single well-defined mark, there would be no need of Direction 1.

the load of either platform is to be removed or much changed, the other platform should first be pushed down as far as it will go.

5. If the balance has by any accident suffered a disturbing shock, the bearings should be worked back into good condition by pressing firmly down on both platforms and then rocking them up and down several times.

6. There may be small differences between large weights which are marked alike. Therefore, in any case where a small change in the weight of some heavy object is looked for, as in Exercise 34 (of the list given in Chapter X.), on the Density of the Air, or in Exercise 52, be careful to use the same large weights in all the weighings.

7. Take great care to let no mercury come into contact with brass weights. Mercury is often carried as a film on the fingers, and is usually to be found in small beads on the table-top.

8. Always replace the weights in their proper holders when the weighings are finished.

Measurement of the Expansion of Air.

I have been from the first much interested in those two exercises of the Harvard *Descriptive List* which undertake to measure the two coefficients of expansion of air, and have been somewhat disappointed at the comparative neglect of them in the schools. I must confess, however, that they present some difficulties, and that the results obtained from them by my own classes have not been always satisfactory. And yet, I am so fully persuaded of the importance of these exercises, that, instead of giving them up, I have lately revised pretty thoroughly their details, in the hope of making the work easier and the results better.

The glass tube which is to contain the imprisoned column of dry air should be at the start about 45 cm. long. The diameter of its bore should be about 0.15 cm.; for, if it is much smaller than this, the capillary effect on the mercury column which confines the air, which effect will rarely be quite the same at both ends of the column, may prove troublesome, and, if it is much larger than this, the mercury column is likely to break, especially in transportation.

Comparative calibration of different parts of the tube is needed for satisfactory measurement of the expansion of air under constant pressure. If tubes could readily be found in which the increase of length of the air column, during heating at constant pressure, would occur in a part of the bore differing not more than 1 per cent in cross-section from the part occupied before expansion, it would hardly be worth while to keep a detailed record of the calibration. But, as this standard of uniformity cannot well be maintained, a record of the calibration should be kept, and each tube in its finished condition should be accompanied by such a record, which may well be in the form indicated by Fig. 7.

Calibration of Tube.

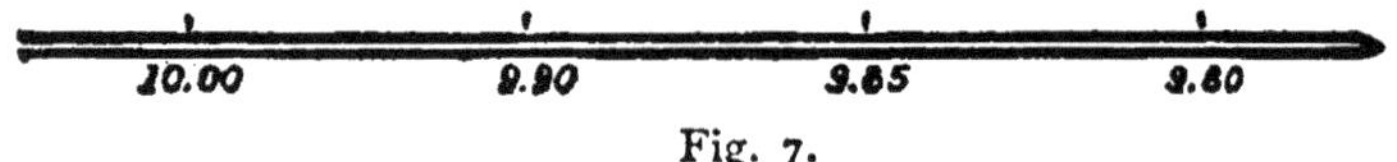

Fig. 7.

This would mean that a certain mercury column was 9.80 cm. long when its middle was at a point indicated by the dot nearest the now sealed end of the tube; that the same column was 9.85 cm. long when its middle was at the point marked by the next dot, and so on. Expansion, during heating at constant pressure, will in this tube take place mainly or wholly in the section between the point marked 9.90 and that marked 10.00. The mean length of the calibrating column in the three right hand sections being 9.85, the variation of bore of the tube will, if neglected, cause an error of about 10 parts in 1000, or 1 per cent, in the value of the coefficient of expansion calculated from the observations at constant pressure. If the calibrating is well conducted and recorded, still less uniform tubes than the one here imagined may well be used, proper correction being made for the variation of bore.

The outer diameter of the tube should be about 0.6 cm. This will insure sufficient strength and will make a good but not too tight fit where the tube is pushed through a 0.5-cm. hole in a rubber stopper, as it will be in use after completion. The companion glass tube, which will be connected with the one already described and will contain the outer end of the mercury column, will be called the outer tube or the open tube. This outer tube should be at least 50 cm. long, but in other dimensions as like as may be to the inner tube which is to contain the imprisoned air. The outer and the inner tubes are to be connected by a piece of antimony rubber tubing about 6 cm. long, 0.3 cm. in diameter of bore, and 0.8 cm. in outside diameter. Into this tube each of the glass tubes should extend about 1.5 cm., leaving about 3 cm. clear. This arrangement will enable us to place the glass tubes at right angles with each other, as in Fig. 10, without danger of collapsing the rubber connection. The bore of the rubber tube being much larger than the bore of the glass, four times as large, in cross-section, we will suppose, a sort of pocket will exist between the glass tubes, which must be remembered in the operation of filling.

Diameter of Tube.

Connections.

This operation is one of critical importance, for if it is not properly done, insuring dry air for the expansion, the apparatus is worse than useless. It is, of course, not enough merely to make sure that there is no visible moisture in the tubes. Care must be taken that no invisible layer of water, sufficient to produce any considerable pressure of vapour on evaporation, exists within the closed air space after the filling. It is practically impossible to dry out a small-bore tube after one end is sealed. For sure effects, dry air must be drawn through the tube while it is hot, and this process should continue for some little time, several minutes at least. The heating is readily accomplished by means of a steam jacket (Fig. 8), about 36 cm. long, that is, about 4 cm. shorter than the finished sealed tube is to be. The air is well enough dried by making it bubble gently through eight or ten centimeters of sulphuric acid among glass beads.

Care in Drying.

In preparation for filling, take the selected and calibrated tube which is to contain the air and, applying to it, 5 cm. from one end, a slender gas flame (through a burner consisting of a piece of glass tubing 2 or 3 cm. long and about 0.05 cm. in diameter of bore), reduce the diameter to about one third of its original size, making this reduction affect as little as may be of the length of the tube. Place the tube in the heating jacket and connect it with its companion glass tube by means of the rubber tube already described, taking care that there be no visible moisture in any of these tubes when they are put together.[1] Connect at the left, Fig. 8, with an aspirator

Making Connections.

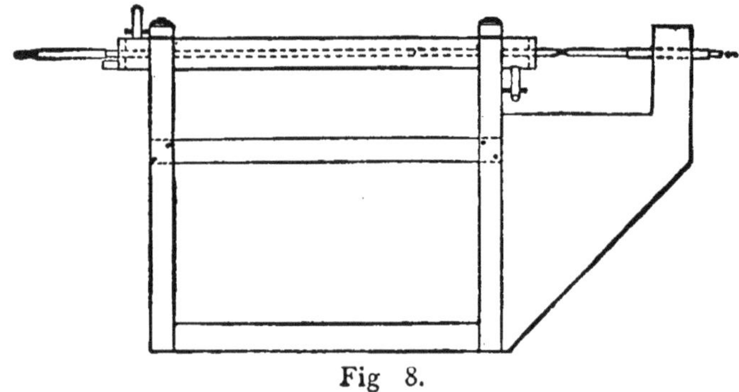

Fig 8.

pump. Connect the drawn-out end of the air-tube, at the right in Fig. 8, with the drying bottle. Especial care should be taken to have the junctions between rubber and glass tight, otherwise steam, escaping from the heating jacket, may be drawn into the tube by way of these junctions.

Let steam flow through the jacket, and draw air through the heated glass tube at the rate, let us say, of 1 cu. cm. per second, to be estimated roughly by means of the bubbles rising from the acid in the drying bottle. The arrangement thus far is indicated by Fig. 8, which shows a

Process of Drying.

[1] It is well to have a tiny wad of cotton in each end of the main tube while it is being put into the heating jacket; for this jacket, as well as the stoppers at its ends, through which the glass tube must extend, is likely to be wet, and without precaution water may thence enter the glass tube.

device for relieving the reduced section of the glass tube from the weight of the connections. The arrow indicates the course of the air coming from the drying bottle.

Maintain this condition of things ten minutes; then disconnect the pump. Apply the small flame once more to the narrowed section of the glass tube until this is melted off and neatly sealed. Allow a little time for the glass to cool after the application of the flame and then withdraw it from the steam jacket, keeping it all the time connected with its companion tube. Of course the heated tube in cooling draws in some air from the outer tube or the rubber connection, but, as this air has been dried, it will do no harm. The undried air of the room will enter the outer tube only. It is probable that the inner tube would acquire no injurious amount of vapour if the operation of introducing the mercury were deferred for some days or even weeks, though, if this delay were to occur, it would be safer to detach the inner glass tube from the rubber connector and seal its open end with paraffin for the time being. Dry air should, in this case, be drawn through the connector and the outer tube before reconnecting for further operations.

Sealing.

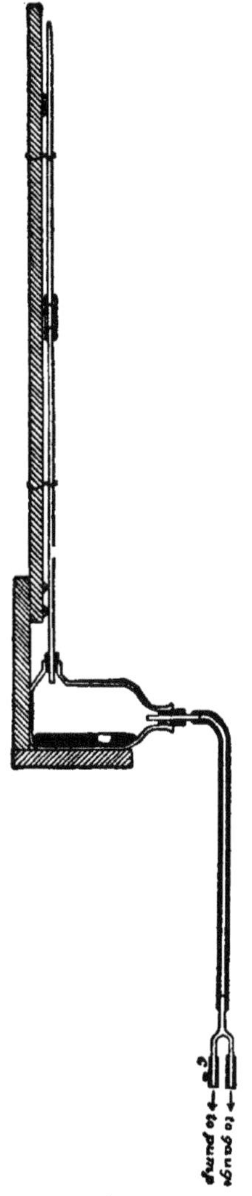

Fig. 9.

The business of putting in the mercury is facilitated by the use of a mercury bottle, or reservoir, having a side opening near its bottom, as in Figs. 10 and 11. The tube to be filled, being connected with this opening, will receive the mercury horizontally, under a pressure which need differ but little from that of the atmosphere, — a condition of things which make it

Introduction of Mercury.

easy to control with considerable nicety the amount of air to be left in the tube. This is an important consideration; for, if too little air remains, small errors of volume become important; and, if too much air remains, expansion at constant pressure may reach beyond the limit of the inner tube.

Adjustment of Air-Column.

Connect, then, the tube, the mercury bottle, a pressure-gauge and the aspirator air-pump, according to the indications of Fig. 9, which represents the bottle lying on its side and does not undertake to show the method by which it is fastened to its support. Before the pump is set in operation, a rough calculation should be

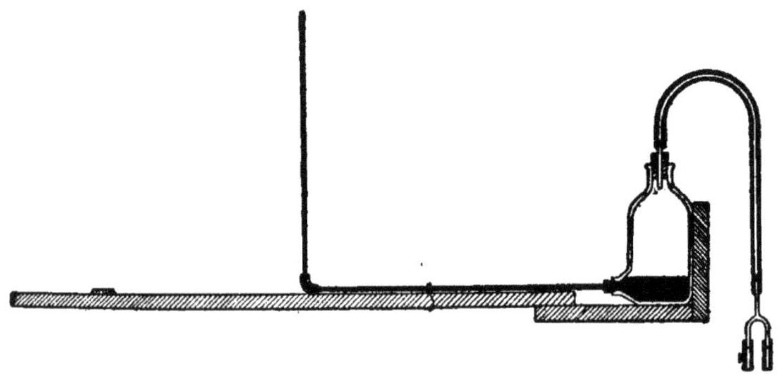

Fig. 10.

made of the degree of exhaustion required. Let us suppose that the closed tube is 40 cm. long, the rubber connection equivalent in capacity to 12 cm. of the glass, the outer tube, of same bore as the sealed tube, 50 cm. long, making a total of 102 cm. We wish to have the imprisoned dry air column, at atmospheric pressure and ordinary temperature, about 26 cm. long. Accordingly, before admitting mercury to the outer tube, we should reduce the air pressure within the tube to 26 / 102 of the barometic pressure, about 19 cm. It is well, however, to make a first trial with a somewhat greater pressure; for it is better to take out too little air at first than too much. When, therefore, the gauge indicates about 21 cm. pressure, close the pinch-cock *c*, disconnect with the pump beyond *c*, tip the bottle up into the position shown by Fig. 10, thus bring the mercury over the end of the tube, lift the sealed tube to a vertical position, as in the figure, so that the mercury entering will leave no air bubbles in the rubber connector, and then open the pinch-cock, letting the air in rather slowly. When equilibrium is reached, lay the sealed tube horizontal, as in Fig. 11, and

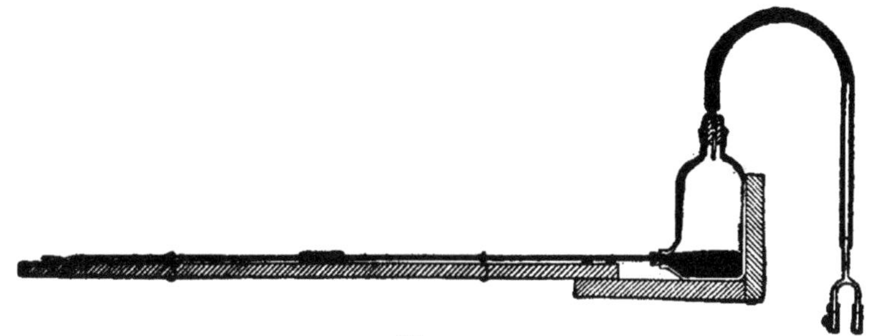

Fig. 11.

measure the length of the imprisoned air-column. If this is more than 27 cm. long, it will be well to take out a little more of the air, for which operation the position of Fig. 9 should be resumed. If the length of the air-column proves to be between 24 cm. and 27 cm. long, it is well enough, and the tube should be disconnected from the bottle, the apparatus being, for this operation, laid on its side, as in Fig. 12, in order to avoid

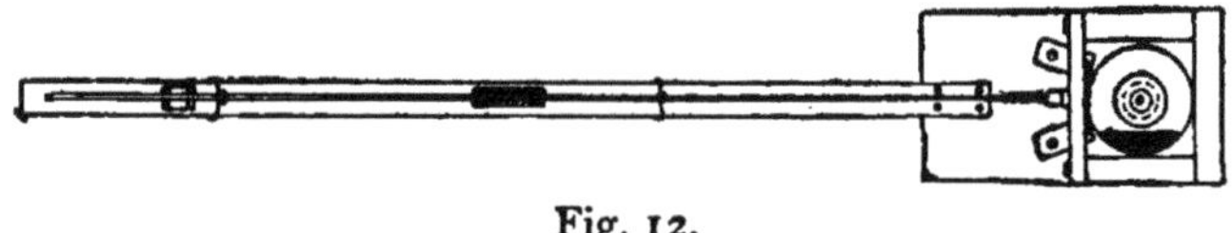

Fig. 12.

spilling mercury. A part of the mercury in the outer tube should now be allowed to run out, so that the column finally left in it, when the whole is laid horizontal, shall be not far from 35 cm. long.[1]

[1] I have lately tested 23 tubes prepared, either by myself or under my supervision, in general accordance with the directions here given. Two of them were bad, giving more than .004 for α, the coefficient of pressure increase. This probably indicated incomplete drying, due, perhaps, to

Calibration, etc., by Manufacturers.

If manufacturers would properly select, calibrate, clean, dry out and seal the tubes, and close one end temporarily with paraffin (see p. 263), it would be an easy matter for teachers to fill them. This division of labour would avoid the danger of breaking, to which the mercury column is subject during transportation.

---

some carelessness in following directions. The others gave for *a*, values ranging from .00356 to .00379, the mean being .00366. Most of the variation was probably due to inequality of capillary effects at the hot and cold ends, respectively, of the mercury column. Each end of this column should be jarred smartly just before readings are made.

CHAPTER V

SCHOOL TEXT-BOOKS OF PHYSICS

REFERENCES.

This list is made up of general text-books intended for school use, most of which include or have reference to directions for a course of laboratory work to be done by the pupils. Laboratory manuals, as distinguished from text-books, special treatises on parts of physics, and general text-books suitable for the use of teachers rather than pupils, will be named later in connection with Chapters VIII. and X.

Avery, E. M. Elementary Physics. New York, Butler, Sheldon, & Co. School Physics. Same author and publisher.

Carhart and Chute. High School Physics. Boston, Allyn & Bacon. 1902. Pp. 433.

Cooley, L. R. C. Student's Manual of Physics. New York, American Book Co.

Crew, Henry. Elements of Physics. London and New York, Mac millan. 1900. Pp. 353.

Gage, A. P. Boston, Ginn & Co.

Elements of Physics. 1898. Pp. 381.

Introduction to Physical Science. 1896. Pp. 374.

Principles of Physics. 1895. Pp. 634.

Gilley, F. M. Principles of Physics. Boston, Allyn & Bacon. 1901. Pp. 552.

Hall, E. H. and Bergen, J. Y. Text-book of Physics. New York, Henry Holt & Co. 1897. Pp. 596.

Henderson, C. H. and Woodhull, J. F. Elements of Physics. New York, D. Appleton & Co. 1900. Pp. 388.

Hoadley, G. A. Brief Course in General Physics. New York, American Book Co. 1900. Pp. 463.

Nichols, E. L. The Outlines of Physics. London and New York, Macmillan. 1897. Pp. 452.

Rowland, H. A. and Ames, J. S. Elements of Physics. New York, American Book Co. 1900. Pp. 252.

Stone, W. A. Experimental Physics. Boston, Ginn & Co. 1897. Pp. 378.

Thwing, C. B. An Elementary Physics. Boston, Sanborn & Co. 1900. Pp. 371.

Wentworth, G. A. and Hill, G. A. A Text-book of Physics. Boston, Ginn & Co. 1898. Pp. 440.

Wright, M. R. Elementary Physics for the Use of Schools and Colleges. London and New York, Longmans, Green & Co. Pp. 256.

We have now considered the natural qualifications of a teacher of physics, his formal professional education and some of the means and habits by which he can keep himself a student and a thinker during the routine practice of his profession. It is now time for us to consider what should be his general method or theory of teaching.

If I had undertaken to write on this subject twenty years ago, it is doubtful whether I should have said anything about laboratory work for the pupils in general of secondary schools. No such work was provided for by the text-books in common use at that time; and the school-teachers of physics, who, as a rule, had never enjoyed laboratory opportunities for the study of the science, and who often were mere conscripts in the service of physics teaching, were not much in the habit of extending their activity beyond the reach and directions of the book in hand.

Retrospect.

Fifteen years ago I should have had to consider the question, strictly a question at that time, whether class laboratory work in the school teaching of physics is practicable or desirable. To-day the question is, What laboratory work shall be done, how much, and in what spirit? Practically, all the American physics text-books written for use in secondary schools nowadays take class laboratory work for granted, although they differ among themselves in the amount and character, or motive, of the exercises which they recommend.

The main purpose of the text-books twenty years ago was to give information. Training of the senses and of the powers of observation and reflection was hardly considered in their construction. They had little more care than dictionaries have for the intellectual processes of their readers. In their one function they probably were, when faithfully studied, fairly successful. For example, Arnott's *Ele*

General Information. Type of Book.

ments of Physics, 1877, recommended to schools by Harvard University in 1881, contained a vast number of interesting statements, most of which would still be accepted as true. It had 873 pages, and, though it was without an index, its table of contents covered 13 pages. It dealt with an enormous variety of particulars. But it did not, I believe, give one problem of any kind for solution by the reader. In fact, I doubt whether there was in the whole book a single interrogation, except those of a rhetorical character. The object and hope of the book seemed to be, so fully to anticipate all needs and questions of the reader that he would never have to do any thinking on matters of physics.

Of course, this book was the extreme of its type, and for that reason I have described it, my object being not so much to show how great the change has been in twenty years, as to make plain what manner of book results from the complete neglect of the training function. In spite of the fact that many a middle-aged man remembers such books with a kind of fondness, and loyally declares his indebtedness to them, a general recognition of their superficial and unsatisfactory character has gradually. retired them from use. Books appealing more to the reason of the pupil, treating him as a growing thing, to be fed, developed, and trained, rather than as a receptacle to be filled, have taken their place.

General Change of View and Method.

It is not in physics alone, or in natural science alone, that a change like this has occurred. A general movement of the same kind was characteristic of the latter part of the nineteenth century in all fields of educational activity in America. One of the first text-books of physics, if not the very first, to show this movement was the *Elements of Physics*, written by Mr. A. P. Gage of the Boston English High School, which appeared in 1882. This book carried on its cover the exhortation, "Read Nature in the Language of Experiment," and it described many experiments to be performed by the pupils. The Author's Preface began thus: "In his report for the year 1881, Mr. E. P. Seaver,

Superintendent of the Public Schools of Boston, says: 'It is a cardinal principle in modern pedagogy that the mind gains a real and adequate knowledge of things only in the presence of the things themselves. Hence, the first step in all good teaching is an appeal to the observing powers,'" etc. That is, laboratory work, the laboratory method of study for pupils of the high school age, was in the air, so much so that its most enthusiastic advocates did not trouble themselves to argue for the desirability, but only for practicability, of such a feature in the school course.

Gage's book soon came into wide use, and it must have exerted a great influence on the methods of instruction in physics in the United States. Meanwhile there were other much-used text-books, good in their way, which I do not mention here by name, because they were not particularly influential in pushing forward that revolution of practice the course of which I am roughly tracing.

Harvard Action on Physics for Admission.

An act of importance in this history was the publication, in 1886, of the following statement, as a definition of one of the alternative requirements in physical science for admission to Harvard College: "A course of experiments in the subjects of mechanics, sound, light, heat, and electricity, not less than forty in number, actually performed at school by the pupil. These experiments may be selected from A. M. Worthington's *Physical Laboratory Practice* (Rivingstons, London, 1886), or from the *New Physics*, by John Trowbridge (Appleton & Co., New York), or from any similar laboratory manual." This was supplemented by the further statements, "In the second of the alternatives in elementary physics, . . . the candidate will be required to pass both a written and a laboratory examination. The written examination will be directed to testing the candidate's knowledge of experiments and experimenting, as well as his knowledge of the principles and results" of the science. "The laboratory examination will be directed to testing his skill in experimenting. At the hour of the written examination the candidate will be required to hand in the original note-book in which he recorded the steps and results of the experiments which he performed at school," etc. "Most pupils will need lectures or other oral explanations in addition to the descriptions given in the laboratory manuals. When it is impossible to provide lectures, an additional text-book made from a different standpoint will be found of advantage."

From Arnott's *Physics* (*Harvard Requirements*, 1881) to this statement is a very wide swing of the educational pendulum, — too wide, indeed. Chemists were more influential than physicists in framing the new science requirements, and they made the mistake of treating the physics just as they treated the chemistry. The emphasis was all put on "experiments and experimenting," the other work of the proposed requirement being mentioned later and almost incidentally. The written examination was to be directed in part to experiments and experimenting, while the laboratory examination must be devoted exclusively to experimenting. Undoubtedly this was too great a reaction from the methods of the preceding decade.

Change Too Great.

The latitude of choice, in laboratory work, permitted by the letter of the new requirement, forty experiments chosen at will from either of two very unlike manuals, "or from any similar laboratory manual," made it necessary for the College to get out a *Descriptive List* of acceptable experiments, that is, a pamphlet giving detailed directions for the performance of the prescribed number of laboratory exercises. This List soon came into very common use within the especial sphere of Harvard influence, and presently a number of text-books appeared which were prompted, directly or indirectly, by the course of laboratory work laid down in this pamphlet. These books, though differing considerably among themselves, constituted a strongly marked type, almost the antipodes of the Arnott's *Elements*, which had not long preceded them.

Influence of Harvard "Descriptive, List."

But some years of use brought to general recognition the already noted defect of the Harvard requirement, and of the text-books corresponding to it, the over-emphasis on exacting laboratory work, and the consequent lack of opportunity for sufficiently varied instruction. Accordingly, in 1897, the Harvard Catalogue gave a new statement of the requirement in Elementary Physics, from which statement the following extracts are taken: "*Elementary Physics*, — A course of study dealing with the leading elementary facts and principles of physics, with quantitative laboratory work by the pupil.

A Revision.

"The instruction given in this course should include qualitative lecture-room experiments, and should direct especial attention to the illustrations and applications of physical laws to be found in every-day life."

"The pupil's laboratory work should give practice in the observation and explanation of physical phenomena, some familiarity with methods of measurement, and some training of the hand and the eye in the direction of precision and skill. It should also be regarded as a means of fixing in the mind of the pupil a considerable variety of facts and principles."

The *Descriptive List* was re-written to fit the new requirement. But such text-books as have been written with especial reference to this List, even in its revised form, continue to be a rather marked type, distinguished by their great attention to laboratory work, by their detailed description of such work, and by the intimate relation which they maintain between it and work of other kinds.

In most parts of the country the change to laboratory methods of instruction was less rapid and less sweeping than in those places where the influence of Harvard was predominant; but a movement of the same general character has been widespread, many teachers in many places helping it on, each in his own way. Ann Arbor, through the books written by men connected with her educational institutions, early became and has remained a centre of great influence.

Other Influences. Ann Arbor.

It is quite probable, however, that, even to the present time, the most successful text-books, from the point of view of numbers sold, have been those making little of laboratory work, following, rather than leading, the change of opinion and practice. Books still appear in which the treatment of laboratory work seems rather perfunctory. It must be said in favour of such books that for cyclopædic purposes or any hasty use they are more convenient than those in which the laboratory work is more prominent and more closely amalgamated with the other parts.

Conservative Books.

The ideal text-book, which is to satisfy all kinds of teachers and all kinds of schools and sweep all competitors from the field, has, apparently, not yet appeared, in spite of the many efforts which are constantly being made to produce it. There seems to be, for the young teacher, no escape, at present, from the task of studying the conditions of his own school, studying the characteristics of a number of text-books, and then making his choice according to his own best judgment, choosing more than one book if this is practicable. In a few years he will be writing a book of his own, which may prove to be *the* book.

No All Sufficient Book.

In the next chapter I shall discuss and illustrate various theories, or views, of laboratory work which appear to have guided the authors of current school text-books of physics.

CHAPTER VI

DISCOVERY, VERIFICATION, OR INQUIRY?

REFERENCES.

Armstrong, W. E. The Heuristic[1] Method of Teaching. Vol. 2 of Special Reports on Educational Subjects. Department of Education of the English Government. Pp. 389–413.

Cajori. The Pedagogic Value of the History of Physics. SCHOOL REVIEW. May, 1899. Pp. 278–285.

Fouillée, A. Education From a National Standpoint. London, E. Arnold. New York, D. Appleton & Co. 1892. Pp. 332. Chap. II.

MacGregor, J. G. Knowledge-Making. NATURE. Dec. 14, 1899.

The Prefaces of all American school-books of physics published during the last ten years.

"Inductive" and "Deductive."

YEARS ago, when I was a novice in the laying out of a course of physics for schools, an eminent professor declared to me with much impressiveness, "There are two ways of teaching science, the inductive method and the deductive method. The inductive method is the only one that has any business in the schools." Another, still more eminent, authority, when in his presence I happened to use the phrase, "the verification of Boyle's law," exclaimed with emphasis, "That is the very attitude we want to avoid, the attitude of verification."

In the light of all my observation and experience, I am now of the opinion that the first maxim was a very bad one, and the second a very good one. The first is a harmful exaggeration of the truth of the second.

It is certainly unwise to take the position that the pupil must be brought to the acceptance of every truth by the roundabout

[1] See also the last chapter of this book.

method of proceeding from axioms along a continuous chain of demonstration, which he must put together link by link. It is well to follow this course in geometry; it is well to follow it often in physics, partly for the mental training it gives, and partly for the light it throws on the methods by which others have sought out the truths of the science; but it is absurd to say that the young pupil should be permitted to take nothing at second-hand. Youth is too short, and life is too short, for such a doctrine.

It is probable, indeed, that no teacher and no writer of text-books would profess to hold this doctrine without limitations. But, nevertheless, the "inductive method," or the method of discovery, is often overworked, with the result that it must break down or be continued as a mere pretence, most harmful to the intellectual morals of all who are parties to it.

Abuses of Inductive Method.

For example, it is well, of course, to encourage and require the pupil to put down at the end of an experiment what he has got out of it; but it has often seemed to me that the printed order to "infer," which is given at the end of each exercise in some of the ruled and formulated note-books now so much used, is often too large for the occasion and leads the pupil to put down some unqualified law, not justified by the evidence which he presents, or to venture some platitudinous remark where the only real inference is a numerical result.

If my memory is not playing me false, I once saw directions for an experiment in which the pupil was to watch intently for some time a bit of wood lying on a table, and then, after reflection, to write down the inference, *Matter cannot set itself in motion.* This experiment was very consistently followed by another, in which the pupil, after poking the object and duly reflecting, was expected to write down the inference, *Matter can be set in motion by force.* I remember another book of experiments, to be used at home by young boys, which began by the dropping of a stone into a vessel filled with water, and developed the subject and the pupils so rapidly as to ask the latter, on the second page, whether, in their opinion, according to Experiment 4, sunlight is matter.

Such is the method of discovery at its worst. It would be unfortunate for young pupils to get the notion that science has been developed by such methods of observation and inference as those just indicated.

The Method at its Best.

But what of this method at its best? Doubtless at its best it works very well; but what are the conditions necessary for this success? A very competent teacher, who knows the ground thoroughly, and will not delude himself or his pupils with exaggerated notions of their independence and originality in science, and a very small class. The method, sometimes advocated, of teaching children to swim by throwing them into deep water, will surely be fatal to a very large proportion of the unhappy youngsters, unless there is some experienced person with every group of three or four. For a single pupil in the art of swimming a judicious teacher is better than a life-preserver; but, if the teacher must have fifteen or twenty beginners in charge at once, in the name of humanity let him give them something to float with, or keep them very near the shore. Usually, I am sure, the teacher who thinks to let his pupils "find out everything for themselves" will find out for himself that he has somehow got the hardest part of the undertaking. For visible progress must be made, tangible results must be reached; the teacher must somehow bring things to pass, in spite of the vast capacity for going wrong which marks the efforts of the ordinary individual, as it has marked the efforts of the human race, to "find things out."

Art of Discovering General Laws cannot be Taught.

Young people are not averse to games of hunting; but, if the hunting lasts very long without result, most of the participants will fall out, and the game, in school or out, will flag. Most general laws or relations in physics are too difficult for the pupil to seek out. Even when all the necessary data are at hand, the unprompted recognition, the genuine discovery, of the resultant law is either an accident or an inspiration akin to accident. The art of such discovery cannot be taught. But physics is peculiar among the natural sciences in presenting in its quantitative aspect a large number of perfectly definite, comparatively simple, problems, not beyond the understanding or physical capacity of young pupils. With such problems the method of discovery can be followed sincerely and profitably.

Method of Verification.

Let us now consider the method of verification. It is hard to imagine any disposition of mind less scientific than that of one who undertakes an experiment knowing the result to be expected from it and prepared to look so long, and only so long, as may be necessary to attain this result. Better by far to take a statement on faith than to cultivate the habit of hunting for evidence in its favour and shutting one's eyes to inconvenient evidence against it. A trait that has characterized the great masters of science has been the power and habit of sternly searching the evidence for, as well as the evidence against, their own propositions.

Prejudice Perverts Evidence.

The suggestion that pupils whose minds are prejudiced in favour of a certain belief will pervert the evidence of their own senses is sometimes ridiculed or resented; but, unfortunately, I have seen too many instances of such perversion to doubt its prevalence. It is sometimes conscious, sometimes unconscious. Even when conscious, it is frequently quite open and without sense of wrong-doing. "Why should I put down an observation which I know can't be right?" the boy will ask in perfect innocence.

Of course, there are cases in which the conditions of a particular observation are such as to make it peculiarly uncertain, — much more uncertain than the other observations which would naturally be grouped with it, — and in cases like this the observation in question should be rejected, even if it happens to agree with our preconceived notion of what it should be. But when all the observations of the same sort have been made under equally good conditions, so far as these conditions are known, we must not reject some of them because they do not agree with the others, or because they do not agree with our notions of what they ought to be. Yet I have again and again found the tendency to do just this thing, and not always among pupils only.

There is very little to choose between the method of verification at its worst and the pseudo method of discovery. The former says to the pupil, *The fact is so and so ; make observations accordingly.* The latter says, *Make observations; from these discover the fact, which is so and so.* We need something better than either of these methods to justify the expense and work of laboratory courses.

Method of Inquiry.

I would keep the pupil just enough in the dark as to the probable outcome of his experiment, just enough in the attitude of discovery, to leave him unprejudiced in his observations, and then I would insist that his inferences, so far as they profess to be derived from his own seeing, must agree with the record, previously made and unalterable,[1] of these observations. The work may relate to some single phenomenon, fact, or constant, or to some general law; but in any case the experimenter should hold himself in the attitude of genuine inquiry.

Spirit of the Teaching.

This attitude is not necessarily inconsistent with fore-knowledge of what the result sought should be according to the testimony of books. Much depends on the spirit of the teaching. If this is such as to show that no forcing or perversion of observations, no pretence in the reasoning from these observations, will be tolerated, if known, the pupils are likely to be powerfully affected by it. But if the text-book used is one that emphatically states laws or quantities, and then instructs the pupil to "verify" these laws or quantities in the laboratory exercises, it will require strong counter-instruction from the teacher to make these exercises proceed in the

---

[1] Sometimes an inspection of the observations will disclose a perfectly obvious blunder, such as a mistake of ten degrees in reading a thermometer. Of course in such cases the record should be amended, with a note describing the change.

right spirit. It is better to keep young observers out of temptation until they are accustomed to depend on themselves.

Consider, for example, the law of Boyle regarding gases. This law is very important, and the manipulation of the apparatus illustrating it is excellent practice for the pupil. But the law is numerically so very simple that, if it were clearly in the minds of the class before the experiment, and especially if the order to "verify" the law had been issued, some in any large class would be pretty sure to feel their way along, making calculations in advance of records, leaving out undesired millimeters, and attain a result in beautiful accord with the idea which inspired them, but probably not quite in accord with the evidence of the apparatus and their own unperverted senses. On the other hand, the parallelogram of forces is a law so complicated that it is difficult for the pupil, when he is making his observations on the magnitude and directions of the three balanced forces, to see whether these observations will or will not lead to a perfect parallelogram, and there is no harm in letting him know, in advance, just what the law is, provided there is some adequate control and check on the use of these observations.[1] The object of the experiment in this case is to make the pupil realize the meaning of the law, while giving him an opportunity to exercise, and by the final result to test, his skill.

Illustrations.

The teacher and the pupils should know that various so-called laws are not strictly true and that even elementary laboratory work may go far enough to show their fallibility. For example, according to my observation of wooden blocks sliding on a surface of paper, friction is not independent of velocity, but is a little greater at high speed than at low, a conclusion contrary to what I should

Inaccuracy of Some "Laws."

---

[1] My practice is to have the pupil make his observations, that is, draw his lines and record his readings, on a sheet of paper placed beneath the strings through which the forces are applied, and then, with a pin, prick through the significant parts of his diagram into a sheet of paper beneath, which sheet remains in the laboratory as a copy of the record.

have expected. The difference is so slight, however, that a boy who has been told, as boys are sometimes told in advance of experiment, that there is no such difference, is pretty sure not to find it, and that very promptly. A good practice in such a case is to tell the class that the difference, or departure from the so-called law, if there is any, is slight, and that, after each pupil has tried the experiment and recorded his unbiased judgment, the opinion of the majority will be taken and announced. The fact of the inequality noted in such a case as this may not be important, but the power and habit of seeing what there is to see, and not merely what one is told to see, is important.

Another instance in which class observation brings out an interesting fact which the experiment was hardly expected to reveal, is found in the exercise on the position of the image formed by a plane mirror. In writing the directions for such an exercise I had made no mention of the need of making some allowance for the refraction by the glass, which virtually brings the reflecting back surface of the mirror a little forward from the straight line over which it is carefully placed. Examining the work of pupils, however, I have found that this refraction does produce a plainly perceptible effect in many cases. If nearly perpendicular incidence and reflection were used, this virtual moving forward would be equal, nearly, to one-third the thickness of the glass; but, as the incidence used is generally very oblique, the virtual reflecting surface, as found by the point of crossing of the lines of incidence and reflection, may be nearer to the front surface than to the back surface of the mirror. It would be unfortunate to blink out of sight a fact like this, even though its complete discussion may have to be postponed for a time.

I have already alluded to cases in which the laws to be illustrated or revealed by the experiment are of such a nature that the work of no one pupil is sufficient for the general purpose, and combination or comparison of results becomes necessary.

One of the best illustrations of this class of exercises is furnished by experiments on the deflection of rods of various dimensions. My habit in dealing with this matter is the following: Each member of the class has at his command two white pine rods, as nearly alike in grain as may well be, each a little more than a meter long, one being 1 cm. square in cross-section, the other 2 cm. by 1 cm. Each rod is studied under various loads, and for each rod and each condition of trial the mean deflection per 1-gm. load is found and reported by the student to whom that particular rod is assigned. The results are then grouped under headings as follows:

Pooling of Observations.

Illustration: Laws of Bending.

1st Case, rod no. 1 (1 cm. square), supports 100 cm. apart.

2d Case, rod no. 1, supports 50 cm. apart.

3d Case, rod no. 2 (2 cm. by 1 cm.) on broad side, supports 100 cm. apart.

4th Case, rod no. 2 on edge, supports 100 cm. apart.

The results found for any one case by different members of the class will differ greatly, partly because the various rods differ somewhat in quality and, very slightly, in dimensions, partly because the work of observation is not first-rate, partly because of numerical errors in the computations. Some reports are so wild, from evident misapprehension of the problem, or from gross numerical errors, like misplacement of the decimal point, that they must be rejected; but it will be seen from the numbers given below that a very liberal standard of admissibility is applied, and that the range of numerical values grouped together is large. Nevertheless, and this is a very instructive lesson to students, general results of considerable accuracy can be worked out from a great mass of individually inaccurate data, provided the inaccuracies are of the accidental sort, so that errors in one direction may, in the general average, be eliminated, or offset, by errors in the opposite direction.

DEFLECTIONS PER ONE GRAM OF LOAD.

	1st Case.	2d Case.	3d Case.	4th Case.
	0.000915 cm.	0.000115 cm.	0.000620 cm.	0.000135 cm.
	800	106	434	109
	1090	140	725	123
	1090	140	700	123
	957	129	587	130
	1050	136	418	111
	1010	135	465	125
	884	103	470	116
	895	120	463	120
	920	120	445	120
	951	124	597	143
	914	109	502	129
	1220	135	505	120
	870	120	520	140
	1050	145	450	120
	1150	160	420	105
	950	147	596	132
	1170	140	435	108
	1140	153	463	123
	1330	130	426	111
Mean	0.001018 cm.	0.000130 cm.	0.000512 cm.	0.000122 cm.

In comparing case 1 with case 2 we get the effect of doubling the length, other things being equal. We see at once that deflection increases with increase of length; but is deflection proportional to length? Is the rule $D \propto L$? Evidently not; the deflection increases in far greater proportion than the length.

Search for the Laws.

Is the rule $D \propto L^2$? This would require the deflection of the 1st case to be only four times that of the 2d case, and it is more than that.

Is the rule $D \propto L^3$? This would require the first deflection to be 8 times the second. In fact, it is 7.8 times the 2d. This is very fair agreement, and shows that the formula last written comes pretty near, at least, expressing the experimental facts of the case. That is all we can claim for the testimony, that it points to the law $D \propto L^3$, which more careful and extensive experiments by others have shown to be very nearly correct.

A comparison of cases 1 and 3 shows the effect of doubling

the width, other things remaining unchanged. The ratio of deflection 1 to deflection 3 is 1.99, indicating the otherwise known law $D \propto 1 / W$.

Comparison of cases 1 and 4 shows the effect of doubling the thickness, other things remaining unchanged. The ratio of deflection 1 to deflection 4 is 8.34, which is in tolerable accord with the accepted law $D \propto 1 / T^3$.

Grouping these various proportions into one general formula, and introducing also the already known rule $D \propto P$, where P is the load, we get

$$D \propto \frac{P \times L^3}{W \times T^3}.$$

This derivation of formulas I do not expect the students to carry through by themselves; for, as a rule, no one student has from his own observations sufficiently reliable data to make the whole discussion satisfactory, and, moreover, the ordinary student could not reasonably be expected to conduct such a discussion without assistance. I put the tabulated individual results on the blackboard, and go through, in a lecture, the course of reasoning indicated by what has just been given. The numbers shown above were reported by twenty members of a class in the year 1900–1901, and were used by me in class substantially as I have used them here.

Another Illustration of Pooling.

The comparison of masses by the acceleration test is another case in which, owing to experimental difficulties, a general aggregation of results is desirable. In this exercise two cars, one loaded with iron, the other with lead, are set in motion by the equal pulls of like tubes of india-rubber, equally stretched, along inclines so adjusted as to neutralize friction. When apparent equality of acceleration has been attained, by varying one or the other load, the cars, each with its contents, are separately weighed. Students work in groups, usually of three, in this undertaking, and record, as their results, the weighings. In 1900–1901 twenty such groups reported the following numbers:

	1st Case.		2d Case.		3d Case.	
	With Iron.	With Lead.	With Iron.	With Lead.	With Iron.	With Lead.
	1460	1410	1240	1255	1010	1050
	1500	1490	1270	1290	1040	1090
	1500	1590	1370	1330	1160	1170
	1490	1450	1260	1240	1040	1030
	1480	1490	1260	1240	1020	1010
	1540	1500	1310	1240	1080	1100
	1550	1450	1320	1290	1090	1090
	1490	1520	1260	1300	1040	1090
	1460	1480	1256	1270	1040	1040
	1460	1430	1240	1220	1010	1010
	1490	1430	1270	1242	1040	1020
	1530	1500	1330	1340	1072	1094
	1540	1520	1350	1320	1080	1080
	1460	1500	1260	1280	1050	1030
	1530	1560	1330	1340	1072	1094
	1500	1480	1270	1300	1030	1040
	1460	1430	1240	1210	1010	1010
	1490	1460	1260	1280	1030	1082
	1480	1500	1250	1240	1020	1020
	1498	1480	1272	1292	1046	1063
Mean	1495 gm.	1483 gm.	1280 gm.	1276 gm.	1044 gm.	1061 gm.
Ratio	1.008		1.003		0.984	

The mean ratio of "with iron" to "with lead," as found from these numbers, is 0.9997. Of course, there is something of luck in the very close approach of this final ratio to 1.

Similar luck is not evident in the figures given in the next table, which shows the results obtained by ten different groups of students, usually four in a group, from exercises 37 and 38 of the Harvard *Descriptive List of Elementary Exercises in Physics*. These experiments deal with action and reaction, and undertake to compare, in terms of an arbitrary unit, the total momentum of two balls before collision with their total momentum after collision.

Another Illustration: Action and Reaction.

In case 1, the smaller ball strikes the larger at rest; in case 2, the larger ball strikes the smaller at rest; in case 3, the two balls meet, each being in motion; in case 4, the smaller ball, encircled by a belt of putty to make the collision inelastic, is struck, at rest, by the larger ball.

The momentum of each ball before collision is estimated from its mass and the horizontal distance it has swung, as a pendulum, to the collision, which occurs when each ball is in its position of rest. The momentum of each ball after collision is estimated from its mass and the distance it swings after collision. As friction of the air does produce a perceptible effect, even in a single swing, this method of estimation gives a slightly too great numerical value for the momentum of each ball before collision, and a slightly too small value for the momentum of each after collision. This defect plainly shows in the results here put down, which were reported by a class under my instruction in 1900–1901.

	1st Case. Momentum.		2d Case. Momentum.		3d Case. Momentum.		4th Case. Momentum.	
	Before.	After.	Before.	After.	Before.	After.	Before.	After.
	1485	1497	2799	3005	2058	2207	5598	5378
	1554	1558	2725	2853	1948	1762	5450	5069
	1516	1282	2447	2727	1989	2015	5495	5455
	1503	1518	1826	1440	1074	1180	3652	3837
	1569	1598	2689	2637	1904	2054	5378	5534
	1515	1458	2745	2626	1988	1729	5490	5094
	1500	1683	1800	1895	1061	1129	3621	3627
	1449	1448	2717	2512	1993	1992	5435	5238
	1554	1558	2725	2852	1948	1762	5449	5069
	1475	1250	1921	1883	1201	1112	3817	3706
Mean	1512	1485	2439	2343	1716	1694	4939	4803
Ratio	1.011		1.041		1.013		1.028	

The mean ratio is 1.023. The air friction, already mentioned, would go far toward explaining the departure of this ratio from unity.

Density of Air Experiment.

A very troublesome exercise, from the point of view of accuracy of results, is that in which an attempt is made to find the density of air at atmospheric pressure by noting the loss of weight of a large bottle of known capacity from which a measured fraction of the air is pumped out, the weighing being done on the familiar platform balance mentioned in Chapter IV. Some teachers maintain that it is not worth while to try to do this experiment, unless a more accurate balance than this one is used; and I must admit that my students, usually working in pairs in this exercise, get some very wild results from their observations.

The time allowed for the actual performance of the exercise is about an hour and a half, and in this time each pair of students is expected to go through the experiment three times. Very careful instructions are given as to the use of the balance.

The following results are obtained from data reported by members of a class working under my supervision in the fall of 1901. I have taken twenty consecutive reports from the record-book, passing over only such as were incomplete or were duplicates of others taken. The calculations of results I have made myself, as my main object at this moment is to show how good or how bad data students get with the apparatus at their command in this exercise. The capacity of the bottle used was about 1900 cu. cm. in all cases. The barometric pressure was about 76.7. In most cases the pumps took out about 90 per cent of the air. It appears that in most cases the pressure-gauge was read with tolerable accuracy, though there is one case of apparently large error in this operation, so that most of the inaccuracy of the results is to be charged to incorrect weighing. The results are as follows:

Density of air in the neighbourhood of 20° C. under a pressure of about 76.7 cm. of mercury:

.00125	[.00047]	.00129	.00119
.00123	.00117	[.00059]	[.00456]
.00115	.00102	.00120	.00083
.00140	[.00174]	.00084	.00104
.00130	.00115	.00110	.00133

I have bracketed four of these quantities, because I think the badness of these four is not properly chargeable to the poor performance of the balance. The [.00174] is obtained from data which appear to be affected by a very large pressure-gauge error. The values [.00047] and [.00059] are from data in which an error of 1 gm. or more is apparently made in the weighing, and the value [.00456] involves an error of about

5 gms. Errors of such magnitudes cannot fairly be laid to the balance. They are blunders, such as might be made with a much better balance. The variation among the remaining values is large, but not, in my opinion, so large as to destroy the value of the exercise, which is an instructive one. The bracketed results being omitted, the mean of the values here given is .00116–, which is about 5% low.

I have intimated that the numbers given in the preceding tables are culled, to some extent, from the reports handed in by students, many reports being so defective as to show that the makers have failed to understand the experiment or have blundered seriously in their calculations. The practice, which I commonly follow, of leaving the class quite uninstructed as to the magnitude of the numerical result to be expected in any case, has this disadvantage, if disadvantage it be, that the student frequently does not know, before he hands in his result, whether it is right or absurdly wrong. This leaves the boy free to make all the mathematical errors of which he is capable; and the number and variety of these which he can put into a simple calculation, especially if it involves a trifle of algebra, is the despair of the teacher, — I cannot say the wonder of the teacher, for the phenomenon, remarkable as it is, soon fails to excite surprise.

Pupils' Blunders.

I have sometimes supposed myself to be afflicted in a peculiar degree by this kind of shortcoming on the part of my students, who, in that one of my classes which has to do with such work as we have been discussing, are for the most part youths who have entered college with a "condition" in physics; that is, they are a picked class, selected by this criterion, that they have been unable or unwilling to learn physics in school. But I find upon inquiry that other teachers, not only in this country but in England also, report a similar weakness in their pupils. Mathematical feebleness and fallibility are the birthright of no small part of every class beginning physics. The only question is, what to do about it.

My own practice, which I do not recommend for schools, is

to put my students on their own responsibility, and require them to stand or fall by the first report they make on an exercise. If a student wishes to repeat an exercise, or repeat a calculation, after finding that his first report is unsatisfactory, he is usually permitted to do so, but with the understanding that a good second report is not to be taken at the same value as a good first report; and repetitions under these conditions are rather infrequent. Thus the full measure of his shortcomings is often not realized by a heedless student until the half yearly day of reckoning comes, and then he is likely to be woefully surprised at the fix in which he finds himself. If I were teaching in a school with pupils some years younger, I should no doubt make a practice of requiring them to repeat, and improve on, work badly done.

Repetition of Exercises.

CHAPTER VII

THE TECHNIQUE OF LABORATORY MANAGEMENT

REFERENCES.

Eastern Association of Physics Teachers. Report on Methods of Instruction in Physics in Secondary Schools. 1900.

Strong, E. A. Physics in The High Schools of Michigan. SCHOOL REVIEW. April, 1899. Pp. 242–245.

Threlfall, R. On Laboratory Arts. London and New York, Macmillan. 1898. Pp. 338.

IN the preceding chapter we have discussed the spirit and object of laboratory work. We have now to consider what may be called the technique of laboratory management.

Report on Methods.

One of the most important questions to be considered under the heading of this chapter is whether a class should be taken through its laboratory work with an even front, all the members of any laboratory section doing the same experiment at the same time, or whether a more open and irregular formation of the forces engaged should prevail. On this question the Report on Methods of Instruction, by a Committee of the Eastern Association[1] of Physics Teachers, issued in 1900, has something to say.

[1] The geographical range of inquiry on which this report is based was much wider than the title of the association would indicate, as the following quotation from the report will show:

"Responses to this circular were received from one hundred and seventy-nine (179) teachers of physics, representing geographically twenty-five (25) States and territories, besides the District of Columbia."

Although I shall have occasion to criticise the report in one or two particulars, I regard it as a valuable document, most of its recommendations being, in my opinion, excellent. The very fact of such an inquiry and discussion as this report represents is a most encouraging indication of the zeal and intelligence now common among teachers of physics. A somewhat similar enterprise was conducted some years ago by an association of Colorado teachers.

Among the propositions set forth by the committee in its introductory circular, sent out for comment and discussion, was the following: "(e) With large divisions, it is economy of time and energy for all the pupils to be at work simultaneously upon the same problem [laboratory exercise] whenever the character of the work will permit."

After the replies had been received, the committee reported as follows: "Simultaneous work upon the same problem by all pupils in the laboratory is not to be recommended as a rule. When, however, the laboratory divisions are much too large for the teaching force engaged, this seems to be the only practicable plan, although its educational value is questionable, and the great expense caused by the necessary duplication of apparatus would prevent its adoption in many schools, and preclude its application to problems with which costly apparatus must be employed."

Even Front?

The reasons for this change of position on the part of the committee, so far as they are expressly set forth in the report, are contained in the following passage: "The proposition" (e) "was assented to with much hesitancy by many, although there were some who seemed to think that such an arrangement would work well; in fact, those who have large divisions generally state that this is their usual manner of working. We note a few of the remarks upon this question: 'I think better results may be obtained with the whole class working on the same problem.' 'The character of the work and the difference existing among pupils will never permit its efficient application.' 'Lack of apparatus would forbid in most high schools.' 'This method is a very poor method and should be adopted only as a last resort.' 'Works toward *mechanical results in California.*' 'Too much like "nickel-in-the-slot" work!'"

This statement of reasons seems to me far from conclusive, and I cannot help thinking that the committee was hasty in admitting, as it seems to do, that the simultaneous method is only a last resort for teachers overborne by too large sections.

Why is the "educational value" of the method "questionable"? The essence of the method is, the same sequence of exercises for all pupils and opportunity for convenient, economical and timely discussion of these exercises by the teacher and the class. Do considerations of "educational value" require a sequence especially adapted to each pupil? Do they require that the explanations and discussions which should accompany every exercise shall be repeated between the teacher and each single pupil of a squad of fifteen, the limit recommended by the committee for the size of a laboratory division? What is more wasteful of the teacher's time, more cruelly exhausting of his nervous energy, than the constant and needless repetition of oral instructions and explanations?

Wastefulness of Irregular Order.

The only way, so far as I can see, to avoid such a painful labour where the irregular order of progress prevails, is so to arrange the apparatus that the printed or written directions will be sufficient to guide the pupil in its use, without oral instruction. But this will have a tendency to work the exercises into such forms that the pupil cannot go astray therein, the true "nickel-in-the-slot" method.

"Lack of apparatus would forbid [the use of the simultaneous method] in most high schools," is one comment. But is this true? In small high schools the reduplication of apparatus required by this method is moderate, and as for the large high schools, according to the committee, "those who have large divisions generally state that this is their usual manner of working." That is, the thing declared impracticable is done and done habitually. The outlay of money needed to provide a squad of fifteen with the apparatus for elementary laboratory work in the simultaneous method is not formidable. Any school board which hesitates to make it, and yet throws the burden of laboratory instruction upon the teacher, should read again the "Song of the Shirt."

Reduplication of Apparatus.

It is true that a rigid following of the method in question would prevent any one pupil from doing more laboratory work than any other. This might or might not be unfortunate; but it is not necessary to be absolutely strict in the practice of this method. It is comparatively easy to devise supplementary exercises for the more rapid workers, to be introduced as occasion requires. For example, if one boy measures the expansion of brass only, another may measure also the expansion of iron.

Elasticity of Method.

Moreover, it is very good practice for even the best pupil to repeat an experiment, going over it as many times as the length of the laboratory period will permit, watching for variations and studying for improvements in his own work. Finally, there are always numerical problems which may well occupy the attention of the exceptionally rapid worker.

It must be remembered that the method of irregular progress does not necessarily imply that one pupil will, in the end, have had more opportunity and done more work than his less efficient classmate. Adaptation of work to individual capacity is a problem in itself. It can be worked out best, other things being equal, by the teacher who has made the best disposition of his other work, and has thereby conserved his own energy and that of his pupils.

Size of Laboratory Section.

We may next consider what is the proper size of a laboratory division. The Committee Report from which I have been quoting in this chapter declares that "the number of pupils in a laboratory division should be about ten or twelve, and *should not exceed fifteen* for one teacher." This recommendation accords well with the opinion which I have long held and frequently expressed. In a large college class, made up to a very considerable extent of young men who have been over the same course of work once before, I have sections of twenty-five or thirty, in charge of a single assistant, but I am not entirely satisfied with this arrangement.

A certain amount of direct personal oversight and criticism, while the exercise is in progress, is needed by most young pupils. As a rule, the teacher should be able to look at the work of every member of the division during every laboratory period.

Individual or Group Work?

Should the pupils work singly, though simultaneously, or in groups, each group having one set of apparatus?

There are some familiar and important experiments which cannot well be done by a single pupil; there are undeniable advantages in serious and honest consultation between members of the class, in the presence of the apparatus. But, on the whole, I believe that co-operation works badly. There is division of responsibility. One or two members of a group will dominate it. If they happen to be interested and energetic, they will do more than their fair share of the work, leaving the others as spectators; if they are indifferent and lazy, they will impose on the others careless and inaccurate methods. Group work is, according to my experience, and I am compelled to use it to some extent in one of my classes, worrying to the instructor and rather unsatisfactory to the students.

I must admit, however, that working in groups of two is the common practice in one of the most successful and satisfactory school laboratories with which I am acquainted. A great deal depends on the spirit which the teacher is able to inspire in his class.

Length of Laboratory Exercise.

The period for a laboratory exercise should, by general agreement, be twice the length of the ordinary school period, as a rule. There are, however, many experiments such as beginners naturally take, those having to do with specific gravity, for example, which can well be done in a single school period, if the work is well planned. It is my opinion that, if the whole laboratory course is extended through two years, as it is in many schools, the work of the first year may well be done in single school periods, the more difficult and longer experiments being taken in the second year.

It was with this possibility in view that the set of experiments given in the Harvard *Descriptive List* was divided into a First Part and a Second Part. It would be a pity to let the real or supposed impossibility of arranging for double periods prevent the beginning of laboratory work. But for such experiments, or exercises, as most of those given in the Second Part of the list just mentioned double periods are needed.

According to the report which has been referred to so often in this chapter, "Both teacher and pupil should be prepared beforehand for the work to be done in the laboratory. The kind and extent of this preparation should depend upon the character of the exercise, the manner in which it is to be approached, its relation to other work, etc." This merely needs amplification and illustration. The preparation demanded of the teacher is physical as well as mental. He must know the theory of the experiment and must have such knowledge of its actual operation as can be acquired in no other way than by going through it with such apparatus as the pupil is to use. Unless he has done the experiment many times in this way, he should have done it recently. He should make sure that all the apparatus which will be needed by the class is in good condition and in the right place, not only the large things, but also the little things, not only bottles and spring balances, but also thread for suspending the bottles on the balances. Not only Bunsen burners, and boilers, etc., but large stoppers for the tops of the boilers and perhaps small stoppers to close holes in the large stoppers.

The Teacher's Preparation.

Unless these small necessaries are thought of in advance and provided, the work of the class is presently suspended, while the instructor trots excitedly about the laboratory, ransacking closets and drawers in a possibly vain attempt to supply the needed article. This wastes time and demoralizes the class.

Almost equally bad is the habit of shouting tardy explanations and instructions to the section after it has begun work. If the teacher cannot practice foresight, he must not expect his pupils to exercise care.

When a printed manual, giving detailed directions for laboratory work, is used, it is hardly necessary or advisable for the teacher to go completely through the experiments by way of example in the presence of the class, nor should he get into the habit of repeating to the class the

Use of a Manual.

directions of the manual; for such a practice lessens the pupil's feeling of obligation to read the directions carefully and robs him of the discipline which he should get by interpreting these directions for himself. The habit of waiting or asking to be told things that are in plain print before him, is one of the besetting vices of the American youth when he comes to college.

The Preparation of the Pupil.

I have discussed elsewhere (Chapter IV.), at considerable length, the question whether the pupil should be told in advance just what he is expected to find in his experiments, the exact law, if he is looking for a law, the exact value of the numerical constant, if he is looking for such a constant. In my opinion he should not, as a rule, be so informed. But he should know enough about the proposed experiment to enable him to get about it promptly, when the apparatus is placed at his disposal, and to go through it with a good notion of what he is driving at. Accordingly, it is well for the teacher to exhibit the apparatus to the class in advance of the laboratory work, with very brief remarks concerning its use, if it and the printed directions are satisfactory, with more extended comments and directions, if the apparatus or the manual is defective. After such an exposition the pupil can read the directions more intelligently than if he had not seen the apparatus, and he should be expected to read them through before beginning the laboratory work.

Oversight but not Meddling.

During the actual progress of the exercise the teacher should maintain a vigilant oversight and be unsparing of helpful criticism, but he should not meddle and he should not demand impossibilities.

Boys working singly, with well-devised apparatus and well-considered methods, will almost always show an excellent spirit in their laboratory manipulations and not infrequently a commendable degree of ingenuity. Their zeal, however, does not always inspire them to put away their apparatus when they have done with it. For example, failure to replace their weights in proper holders after use of a balance is one of the minor evils with which a teacher has to deal. The pupil should find his apparatus in good condition and should leave it in good condition.

Form of Record.

As to the proper character of the pupil's record of his work, I can hardly give better advice than by making one more quotation from the report now so familiar to readers of this chapter: "The pupil should keep a laboratory note-book which should contain a *concise* statement of:

(a) Problem to be solved.

(b) Method of work.

(c) Apparatus and material used; and, in many cases, a rough sketch of the arrangement of the apparatus.

(d) Necessary formulas and computations.

(e) Observed results, together with such inferences as the pupil may reasonably be expected to make."

Nothing is here expressly said concerning the observations, which are, presumably, to be recorded in connection with the "Method of work." It will probably be necessary for the teacher at the beginning of the course to prescribe pretty fully the form of observation record; but the pupils should become accustomed, as the work goes on, to plan their own arrangement of the facts to be put down. The effort should be to write what is essential, and only what is essential, and all in such a form as to be easily intelligible to the reader. The pupil should try to make such a record as would be most useful to himself years afterward, if he should have the task of taking a class through the same experiments. This may seem to be asking a good deal of young pupils, and of course they will fall short of perfection; but the thing to be impressed on them is that a record should tell a plain tale to people who are not present when the record is made, or who, through lapse of time, have forgotten much of what the record sets forth.

Of course, if a laboratory manual giving detailed directions for the work is used, it is not necessary or profitable to copy all of these directions into a note-book. An abstract of the method should be given in the pupil's own language or indicated by the recorded observations.

Practice of the graphical method of record, by means of the simplest possible drawings, is of very great service; for it requires the pupil really to study his apparatus, and yet, by saving many words, may save his time as well as that of the reader.

An Illustration.

The following example may illustrate some of the precepts which have now been given:

Exercise............ *Feb*............, ...,

STUDY OF THE ZERO-POINT ERROR AND THE BOILING-POINT ERROR OF A CENTIGRADE THERMOMETER.

The general method used was the one given in................................., but the boiler used was different from the one shown in the book, having a top that screwed on and a water-gauge at the side for showing the height of water in the body of the boiler. See Fig. 13.

The thermometer tested was No............ It had a paper scale and was graduated in degrees, the scale extending from 10° below 0° to 110° above.

Reading in ice and water (or snow) before heating = +0.2°.

Reading in steam, as in Fig. 13, = 99.7°.

Reading of barometer = 76.5 cm.

Reading of thermometer in steam as in Fig. 13, but with steam outlet nearly closed, and mercury gauge as in Fig. 14, = 101.1°.

(Real difference of level of the mercury columns about 4.2 cm.; allowance of 0.1 cm. made for pressure of water, about 1.5 cm., in left-hand side of the gauge.)

Reading in steam, with steam escape wide open, but with 0° mark just above top of stopper = 98.7°.

Reading in ice and water after heating = —0.1°.

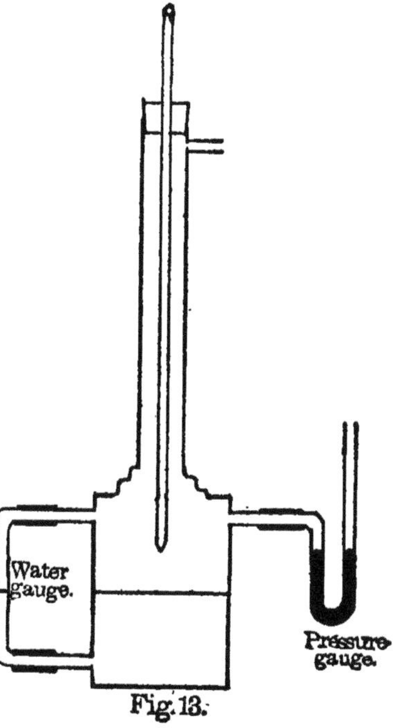

Fig. 13.

Conclusions: The thermometer read about 0.2° too high in ice and water at first, and about 0.1° too low at the last. The reason for this change I do not know, but it appears to have been caused by the heating.

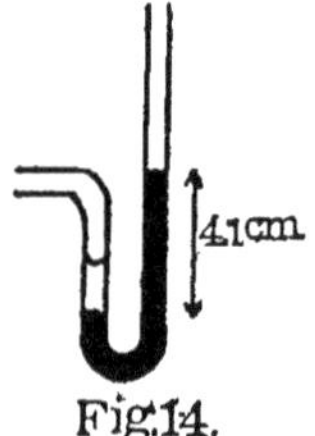

Fig.14.

The thermometer, with bulb and stem in steam, read 99.7° when the barometer read 76.5 cm., 0.5 cm. above standard. According to my observations, a rise of pressure equal to 4.1 cm. of mercury caused a rise of 1.4 degrees in the boiling temperature. This is $(1.4 \div 4.1)$ degrees $= 0.34$ degree, for 1 cm. According to this, if the barometer had read 76 cm., instead of 76.5, the thermometer would have read about (0.34×0.5) degrees, $= 0.17$ degree, lower than it did; that is, $99.7° - 0.17° = 99.5°$, nearly.

Drawing the stem, from 100° to 0°, out of the steam, while the bulb remained in the steam, lowered the reading about 1°.

The next recommendation of the Eastern Association runs as follows: "The laboratory note-book should be written up in the laboratory at the time the work is done. The writing should be in ink so that the original entry cannot be erased." It is doubtful whether, under the conditions imposed by this rule, the average pupil could be expected to complete such a record and discussion as that just given, unless the time allowed for the whole exercise were more than two consecutive school periods. Could not a part of the work, the writing out of conclusions, be done later and elsewhere? If not, I should fear that the observations would be hurried and unsatisfactory.

Time for Record and Calculations.

It is, of course, desirable to discuss the observations and derive the conclusions as promptly as may be. Promptness saves time and contributes toward good results. But independent thinking, even on so simple a problem as calculating what the thermometer reading would be if the barometer reading were changed a certain amount, the calculation to be on the basis of the pupil's own observation of the effect of increased pressure, keeps the ordinary boy thinking for a considerable time, unless he gets very effective help from the teacher or some one else.

It is, perhaps, the "some one else," the possibly injudicious guide and collaborator, that the Eastern Teachers undertake to avoid in their prescription that the record and discussion shall be finished at once and in the laboratory. I must admit that there is danger of a great evil from this side, and I do not see how to avoid it altogether without following the strict and difficult rule which they lay down. This danger, however, is not peculiar to physics. It besets all work not done under the eye of the teacher.

Oversight of Note-Book.

It is true that, in the case of boys applying for admission to college, the college may require the teacher's certificate that the note-book is the candidate's own record of his own laboratory work, but it is evidently unfair to ask that the teacher shall have seen every word of it written, or shall have made sure that every mental operation represented in it is original with the pupil. That the actual laboratory work and handwriting are the pupil's own, the teacher can reasonably be expected to know and to declare; and this declaration assures the college examiners that the candidate has at least gone through the motions of using apparatus and keeping a record. This fact raises a sufficient presumption in his favour to justify the examiners in admitting him to their tests. The mere clerical practice of keeping an orderly note-book is valuable and something to be counted in the candidate's favour.

If teachers find it readily practicable to live up to the rule given by the Eastern Association, well and good. But I fear that many conscientious teachers have expended in the oversight of the record an amount of care, and even of anxiety, quite incommensurate with the attention given to it by the college examiners, and quite unnecessary for the proper function of the note-book.

First Record Should Stand.

The practice of making the record in ink from the start, and in such form that it will not have to be rewritten, is an excellent one. A copied, or rewritten, record is sure to look better than the original, but it is not the main object or virtue of a record to look well. Note-books should not be confounded with copy-books for the practice of penmanship. Moreover, the original notes, if made on loose sheets as they frequently are when copying is intended, are very likely to be misplaced and lost. Taking the notes in one book, and then copying them off in fair form into another book, both books being preserved, is a practice which avoids this difficulty; but it involves labour which is, in my opinion, unnecessary, and, on the whole, unprofitable.

Tentative Work.

The habit, which some teachers follow, of encouraging their pupils to make a hasty first trial of an experiment, the record of which trial is not preserved unless it happens to be satisfactory, seems to me rather objectionable, as it may appear to warrant the practice of culling observations and leaving out those which do not accord well with others.

"Data Slips."

In a college class, made up largely of students who have failed in physics at the admission examination, I am much less regardful of the note-book than I should be in a school with younger pupils. My practice is to require the student to write out and hand in, before he leaves the laboratory, a brief record of his numerical observations in the experiment which he has just performed. These notes he makes on a slip of paper, usually about 3.5×4.5 inches, called a *data slip*, which is soon pasted into a large scrap-book under the student's name, four pages of the book being devoted to his record for the whole course. A similar set of notes, with such amplifications as may be required, usually very few or none, is made by the student in his note-book.

"Result Slips."

The laboratory exercise for each student in this course comes once a week, and one week after he performs any experiment he is required to hand in the worked-out result or conclusion from this experiment on a *result slip*, which is presently pasted by one edge into the scrap-book, just over the corresponding data slip, to which it is like in form and size. The result must be such as the data, which the student has had no opportunity to alter while working out his result, will yield.

By this device, I have at any time during the year a bird's-eye view of what each student in the course has done in the way of laboratory work.

The note-book is examined only twice during the year, and, as it is, in considerable part, a duplicate of the scrap-book record, not very much importance is attached to it. I feel, however, that it would be an injury to the students to dispense altogether with the note-book, the keeping of which helps to keep the lessons of the course in their minds. It will be seen that the practice just described involves some copying, but, as the records are usually brief, this is no great hardship.

As the method of individual laboratory work makes each student's data differ in some particulars from the data of others, the danger of illicit practices in the working out of results is less than one might at first suppose.

Lessons from Laboratory Work.

It is my frequent practice to comment briefly, at the first convenient opportunity, on the character of the results yielded by any given exercise, illustrating my remarks, which are addressed to the class as a whole, by examples from good or bad reports, and exhibiting such tables as are shown in the preceding chapter of this book. True, it is not best to say much to a class in regard to the mental or moral discipline gained from any study. Young people are proverbially averse to sermonizing, and like to feel an immediate motive for what they do. Yet one of the most important objects of laboratory work, properly conducted, is to show the pupil side by side the poor results of poor work, physical or mental, and the good results of good work. There is a convincing tangibility about the results of definite laboratory problems, which is bound to make an impression.

Moreover, the fact that a large number of moderately good results contribute to give at last a very good one, positive and negative casual errors eliminating each other in the long run, though certain constant errors remain to be investigated and discussed, — this is a truth better taught in the concrete than in the abstract, and certain laboratory exercises are peculiarly well adapted to teach the lesson.

In some exercises, which relate to the properties of individual objects, the specific gravity of wooden blocks, for example, or the focal length of lenses, the different members of a class, each having his particular object to work with, may correctly get different results. In such cases, the teacher should mark each of the objects studied and make such a record of its properties that he can readily tell whether any pupil, studying a given object, has or has not done his work well. Some little business ability is needed to make and keep such a record with sufficient accuracy and without an unreasonable amount of labour and worry. But, if it is well planned and vigorously kept up, it will amply repay what it costs.

Teacher's Record of Apparatus.

The obvious fact is that the pupil has a right to know, and that not so very long after the exercise is finished, whether he has or has not done well in any particular task. Without such an assurance he may be unduly discouraged or unduly confident. This being the case, there is no more distressing job for the teacher than that of trying, without an adequate knowledge of the facts, to pass judgment on work done.

It is plain enough that the business of teaching physics by the laboratory method imposes on some one a large amount of purely mechanical labour in the preparation and care of apparatus, to say nothing of its manufacture. It is the duty of the teacher, who needs, of course, to be continually in a state of mental activity and alertness, to save himself all unnecessary physical labour, unless he happens to be so constituted as to enjoy it.

Economy of Teacher's Effort.

As a rule, the teacher should not be expected to make apparatus which he can find ready made to his liking; for he is, presumably, the mental superior and the mechanical inferior of the workman employed by the manufacturer. Even in the handling of the apparatus after it is in the laboratory, there is much purely physical routine labour which a person who can hardly read and write, who is at any rate less highly trained and paid than the teacher, can do quite well

Relief from Manual Labour.

enough. Such is the work of setting out, putting away, and cleaning apparatus, and doing errands and odd jobs of various sorts. The most satisfactory person for this kind of service is one who is entirely satisfied with it, takes pride in it, asks nothing better than to do it, for reasonable pay, all his life. I have known a number of such men, all Irish as it has happened, and could ask for no better service than they have habitually given, after, of course, some painstaking initial training.

But I am aware that such assistants are not usually to be had by school-teachers of physics. Such teachers, within my acquaintance, make excellent use of pupil assistants, or of 'prentice teachers who are willing to work for little pay for a year or two in the hope of acquiring valuable methods and experience. Sometimes these assistants are young women.

Arrangement of Apparatus.

The importance of having some orderly and natural arrangement of apparatus for the moment not in use, instead of a mere haphazard distribution which only one person can remember and which no one can explain, is too obvious to need discussion. There is much in favour of arranging apparatus, as nearly as may be, in the order of its use during the year. This facilitates not only the parading and retirement of apparatus for use, but also its survey with regard to future needs. The assistant should be taught to look weeks, or even months, ahead and give early notice of any lack, in order that the worry and expense of hasty provision at the last may be avoided.

Detailed suggestions for the arrangement and equipment of a laboratory will be given in Chapter XII.

CHAPTER VIII

LECTURES AND RECITATIONS

REFERENCES.

Day, R. E. Numerical Examples in Heat. London and New York, Longmans, Green & Co. 1889. Pp. 176. Gives the answers.

Dolbear, A. E. The Art of Projection. Boston, Lee & Shepard. 1892. Pp. 178.

Jones, D. E. Examples in Physics. London and New York, Macmillan. 1900. Pp. 348.

Matthews, C. P. and Shearer, J. Problems and Questions in Physics. London and New York, Macmillan. 1897. Pp. 247.

Pierce, E. D. Problems in Elementary Physics. New York, Henry Holt & Co. 1896. Pp. 194.

Snyder, W. H. and Palmer, I. O. One Thousand Problems in Physics. Boston, Ginn & Co. 1900. Pp. 142. Gives copies of Harvard admission papers in physics.

Wright, L. Optical Projection. London and New York, Longmans, Green & Co. 1901. Pp. 438.

THERE is no aspect of laboratory work more striking than the poor results which it, when standing alone, yields in written examinations containing problems and illustrations not explicitly occurring in the laboratory.

Laboratory Work not Enough.

The pupil who is dull or lazy at mathematics is very apt to feel that the mere mechanical performance of the experiments set before him should, by any reasonable measure and appreciation of human effort, be enough to save him. He "has been there every time," has even perhaps, "handed in all of the results," right or wrong. Is it not, then, cruel injustice to find him wanting at the last, merely because he "could n't remember all those formulas and things," relating to specific gravity, fluid pressure, levers, the parallelogram of forces, etc. The fact is, that most boys are more inclined to work hard with their hands than with their heads, more willing to handle apparatus than to draw any mental results from their activity. Even older persons have been known to lack the resolution and the intelligence to read all the lessons their own experience has written. Old or young, those who do not need urging and guidance are exceptional.

The laboratory method is a good method, so far as it goes. For most pupils it is essential to a firm understanding, a clear vision, a just perspective. Experience of the senses is the solid ground from which the highest flights of speculation and theory in science begin, and to which they must return, with or without safety to the voyager. But learning by experience is a plodding method, and the student who aspires to any great height or breadth of intellectual reach must not confine himself to it.

Function of Lectures and Recitations.

There are several more or less distinct functions which lectures and recitations, in connection with laboratory work, can perform; the introductory explanation of the laboratory exercises, the derivation and discussion of immediate results from these exercises, the application of these results to problems not given for solution by trial in the laboratory, the introduction of facts and principles not touched by the laboratory work, the discussion of physical phenomena in general. For each of these several purposes the methods of continuous discourse by the teacher, of interchange of question and answer between teacher and pupils, of lecture-table experimentation, will naturally be used in turn.

It is quite impossible to assign to each of these methods its exact relative value, or to state the proportions in which it should be mixed with the others. The teacher perhaps feels surest that he is interesting his class when he is showing experiments, surest that he is getting it ready for an examination when he is conducting or enduring a sort of cross-examination, or "quiz," surest that he is giving a comprehensive view of the field before him, or, rather, setting forth such a view for those to take who can, when he is speaking at length and without interruption.

Interest is almost as essential to the boy in the classroom as to the horse at the watering-trough, and therefore experiments, or at least some form of stimulus other than that yielded by the pupil's bare sense of duty, must be supplied, when the exercise would otherwise be one of unbroken discourse by the teacher. Which of us does not feel a little wearied at the end of an hour's talk by the most esteemed philosopher? Young pupils must not be expected to maintain a continuous mental flight for more than a few minutes at a time. But this warning is probably unnecessary. Few teachers in schools have time to construct set lectures which they would be willing to deliver to their pupils.

On the other hand, the preparation of experiments which can be depended upon to come off at the right time, and with the right effect, is a serious undertaking, if it is to be a frequent one. The tendency is, therefore, I suspect, for the teacher to use very largely, and sometimes to abuse, the recitation or "quiz" method of keeping his class occupied. I use this last phrase advisedly; for the necessity, imposed by the school programme, of keeping a class for the whole of a certain time in a recitation-room, because it would disturb other classes to dismiss this one before the stroke of the bell, must often lead to expedients, more or less conscious, for killing time.

Abuse of the "Quiz."

For this purpose there is no device more convenient or more serviceable than to ask questions, especially questions which the pupils cannot answer with readiness, sometimes questions which are purposely obscure, and so keep up a kind of game of mystification till the hour is over. As an example, let the following serve, a by no means wholly imaginary conversation between an excellent but overworked teacher of science, of all the sciences, and his class in physiology, the subject being the nervous system, and the especial topic, the sensations of a person who has lost a limb:

Teacher. "Now I have heard that sometimes a man whose leg has been cut off will complain of feeling a pain in the toes of the foot that he has lost; he will perhaps feel as if his toes were cramped, and he will ask some one to go and get the leg and — do what?" No answer from the class.

Teacher. "Come now, children, come, speak up — do what? What do you suppose he wants them to do with the leg?"

Pupil. "Bury it."

Teacher. "No, it 's buried already, we will suppose."

Another pupil. "Burn it."

Teacher. "Oh, no! Come, come, children, *what does he want them to do with the leg?*"

Class is silent.

Teacher, as the bell rings, "*Straighten out the toes.*"

Numerical Problems.

The teacher of physics is fortunate above the teachers of most other subjects in having always the legitimate and most salutary resource of numerical problems, to be worked out on the spot, and to be discussed immediately, in the presence of all the class, as soon as they have been done, rightly or wrongly, by a considerable number. Of course this kind of exercise can be overdone, and can be mismanaged otherwise. It is usually necessary to repress one or two bright pupils, who will do the work more quickly than others, and whose superiority in this particular, if not judiciously ignored, will discourage and bring to a standstill the rest of the class. Moreover, the problems to be given should be selected with care. They should be representative, putting into application some important fact or principle, theoretically rather simple and numerically brief. Fortunately, there are good printed collections of problems suitable for the use of beginners in physics, books for which we cannot too gratefully thank the painstaking and public-spirited makers.

Preparation for Lectures.

As to the need of careful preparation for lectures and for lecture-table experiments, there is little call for exhortation. Nearly every one has felt or has seen the melancholy results of the lack of such preparation. But something of possible use may be said in regard to the way in which the teacher can best expend his effort.

In the first place he should consider whether the thing which he proposes is important, and in the next place whether it will produce on the pupil an effect which will justify the labour necessary to prepare and present it.

The teacher sometimes undertakes an experiment without fully realizing its difficulties or the imperfection of the apparatus furnished for its performance, and, having ill success in his first trials, becomes roused to an obstinate effort to make that particular thing work. Such an experience may do no especial harm, may even be, in a way, profitable, if it occurs in a period of leisure when there is time to make experiments, and time to make mistakes; but if it comes shortly before the lecture-hour, it may be disastrous; for men of a certain temperament, when once involved in struggle with difficulties, can think of nothing else for the time being, and if they do come tardily to the conclusion to leave the struggle for a more convenient season, they do so with a sense of defeat that unnerves them for the prompt and confident doing of things that are commonly quite within their powers. Those who have this peculiar form of obstinacy, which may be a source of strength under some conditions, must on unimportant occasions beware of the undertow of their own disposition and keep well above it.

What Experiments are Best.

In the way of lecture-table experiments it is not necessarily the laborious achievement that counts. The little, simple, easily performed, easily seen, but striking, exhibitions of phenomena and illustrations of principle, happily introduced and well executed, are the most profitable things to show, such, for example, as curious hydrostatic effects, various aspects of surface tension, experiments with static electricity, etc.

There are, of course, many desirable experiments which require considerable care at every annual repetition. For example, although I have written out and printed careful directions for the preparation and use of apparatus for the sudden freezing of water, I could not now undertake to make this preparation in half an hour with confidence of success. The method is to boil distilled water for several minutes in a test-tube, then pour oil on its surface, etc. But there is apparently a difference in the adaptability of test-tubes for this use. In some of them the condition of bumpy boiling does not occur, even after ebullition has been maintained for several minutes, and I never feel much hope that sudden freezing will occur where "bumping" has not occurred. For safety, I make ready a number of tubes, three or four, set them all to cool in ice-water, and finally, in the presence of the class, try one after another in the freezing mixture till one has proved a success or till all have proved failures. Similarly, in making preparation for the freezing of water during its own boiling, over sulphuric acid and under the bell-jar of an air-pump, much care is necessary. The pump must be in such condition that it will lower the mercury gauge nearly to 0.45 cm., and a pump which is subject to much and varied uses is not always in this state of effectiveness.

Look Forward a Year.

Whenever a teacher finds that an experiment works well, subjectively and objectively, in itself and on the class, he should leave the apparatus for this experiment in the most secure and convenient condition for use the next year, and should make, if possible, such brief notes in regard to its use as will make all further tentative experimentation with it unnecessary.

Practice.

As a rule, all experiments, whether simple or otherwise, should be tried anew before each exhibition of them; there are so many ways in which they can go wrong. Annual practice in the art of picking up bits of paper by means of an electrified rod of gutta-percha might seem unnecessary care; but I consider it worth while. Sometimes the paper does not come up.

Even when an experiment is perfectly successful for the near-by observer, it is necessary to consider whether it will be apparent to a whole class in a large room. For example, the indications of a gold-leaf electroscope are very likely to be invisible at a comparatively short distance, because of the light of windows reflected from the surface of the glass. Shading the windows, lighting the electrometer by means of a lamp screened from the spectators, and using white paper behind the instrument, makes a great improvement.

For lecture-table experiments with electric currents a galvanometer with vertical index, attached to a needle free to move in a vertical plane, is extremely useful. Fortunately, such instruments are now common in the apparatus market.

Lecture-Room Galvanometers.

For the exhibition of weaker currents, requiring the use of an astatic galvanometer with a mirror, I have found the device illustrated by the following figures very satisfactory. In Fig. 15 (p. 311), *l* is a powerful spiral incandescent lamp (or a Welsbach burner), within an opaque vertical cylinder, *E*, pierced by a small orifice, *o*, through which light goes to the plane galvanometer mirror, *m*, through the converging lens, *c*. After reflection from *m*, the light passes through the plane glass, *g*, to a second plane mirror, *p*, which is held by an adjusting screw, *a*, Fig. 16, passing through the fixed incline, *i*, at such an angle as to send the light to the scale, *s*, which is placed some feet above the galvanometer and is inclined about 45° from the vertical. The various distances are such that an image of *o* is formed on this screen, and this image can easily be seen by a large class in a room but little darkened. The envelope *E*, in which the lamp is placed, should be open at top and bottom, to escape overheating, but above the top there should be a non-reflecting metal screen to absorb the light which, if not arrested, would reach the scale on which the image of *o* is shown.

The especial merit of this arrangement of lamp, lens, mirrors, etc., is that it places the scale directly before the spectators, while leaving the space in front of the galvanometer clear for the operations of the lecturer.

The use of the projecting lantern is now so common, and is so fully described by the publications of the manufacturers, that

I shall not dwell upon it. "Slides" sufficiently good for certain purposes, the exhibition of rough diagrams, for example, can be made without the use of photography by merely scratching the needful lines through the film

Projecting Lantern.

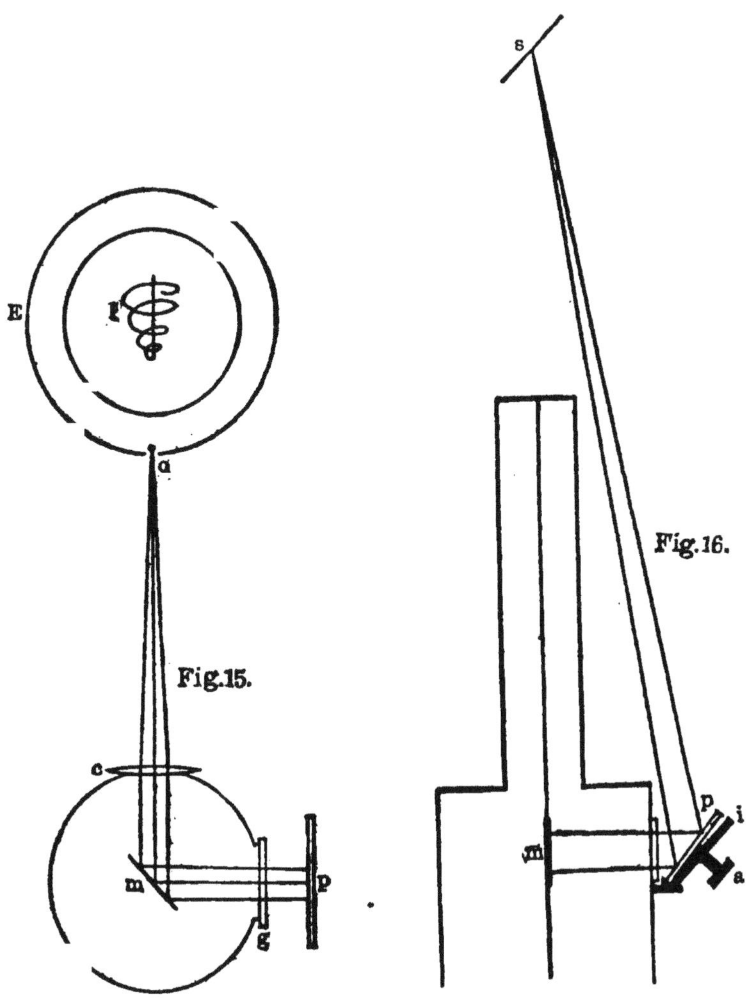

Fig. 15.

Fig. 16.

of an ordinary photographic plate. For extended use of the lantern the usual arrangement of putting it in the rear of the lecture-room, the screen hanging behind the lecture-table, is probably the best; but when its use is a mere incident in a lecture, it is more convenient for the lecturer, who will probably manage the lantern himself, to have it on the lecture-table, the screen being at one side of the room.

An interesting and useful device, not new but perhaps unfamiliar to most people, has for its object the projection of the image of any flat object of suitable size in its natural colours. For example, a picture on the page of a book is illuminated obliquely by means of the condensing lens of the lantern, and the projecting lens of the lantern, detached from its usual position, is used to throw the image of the picture on the screen. Partitions should be used to prevent the escape of too much of the light laterally.

Window-Shades. Opaque roll window-shades, intended for thoroughly darkening a room, are troublesome unless carefully made, with the edges projecting a considerable distance, two inches let us say, if the windows are large, into the window casing on each side. Without this precaution, the shades are likely to bulge inward during a high wind and draw their edges from cover. Such shades should be made to pull down, like ordinary shades, not up, lest the wear and tear on the working cords be too great. If the room is to be darkened but infrequently, light, portable, wooden frames covered with oil-cloth, held in place within or against the window casings by any simple fastening, serve well enough.

Qualitative Experiments. As a rule, qualitative experiments are given to better advantage in the lecture-room, as quantitative experiments are given to better advantage in the laboratory; for the former have generally a spectacular aspect, often sufficiently revealed in a glance at the critical moment, and as easily shown to many spectators as to one, while the latter are more frequently painstaking, prolonged, and comparatively uneventful, requiring also close observation at short range, which cannot be given by the whole class at once.

I used to give as a laboratory exercise a study of the phenomena occurring in the heating and boiling of water, and had contrived for this purpose a small-scale piece of apparatus, which could easily be furnished to each member of a laboratory section. But after some years of trial I came to the conclusion that I had better point out the significant features of the process to a whole class at once than to each member of the class in turn. Accordingly, I now show the experiment, on a comparatively large scale, in the lecture-room.

Similarly, I used to have each student or each pair of students go through certain experiments with a pressure-gauge in water, to illustrate or discover the facts that pressure increases with depth, is independent of direction, etc. Now I do not insist upon this, but show these experiments to groups of students or to a whole class at once. A little contrivance adapts them to the projecting lantern, the index of the gauge being shown on the screen, while the vessel containing the water is in direct view of the spectators. This requires the gauge to be fixed at a certain height and the water in which it is submerged to be moved up and down.

It is well, after showing important[1] experiments like this to a class, to leave the apparatus at the disposal of students, who, in the laboratory, may wish to examine it at short range or to use it for themselves.

[1] I think that I should in some place, here as well as elsewhere, object to that device, still to be seen even in new books, which undertakes to prove the equality of vertical and horizontal pressure at a given depth in water by showing that the water produces the same effect on a vertical mercury column when admitted to the top of it through a horizontal tube, as when admitted through a vertical tube. See *a* and *b* of Fig. 17. This experiment shows that the *vertical* pressure at *a* in the one tube is equal to the vertical pressure at *b*, on the same level, in the other tube; but to say that the vertical pressure at *b* must be the same as the horizontal pressure at the open end of the horizontal part of the tube, is to beg the whole question at issue.

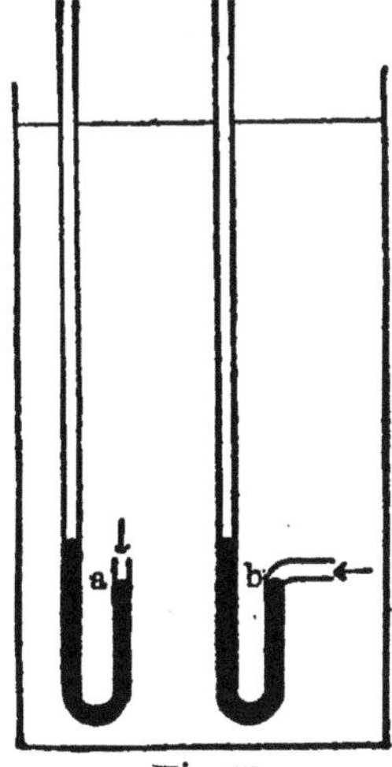

Fig. 17.

It is obvious that certain matters must be treated by lecture and recitation methods, if at all, for the reason that they cannot be brought into the laboratory and put at the disposal of individual pupils. Such are large-scale systems of heating, ventilation, drainage, lighting, transmission of power, etc. These should be presented by the aid of diagrams and verbal explanations, to be followed, if this is practicable under good conditions, by visits to the apparatus itself in position and operation.

Applications of Physics.

It is a question how much instruction on such topics should be undertaken, and this question must be answered with some reference to the local conditions and the general character of the school. It is reasonable that all boys, at least, should acquire a good general understanding of ordinary domestic scientific appliances, and, in their simpler forms, of the steam-engine, telegraph, telephone, dynamo, and motor; but it is easily possible to go too far into details. I can give no better criterion for deciding what things to take and what things to leave untouched, than that which is furnished by the interest and probable future needs, viewed broadly, of the ordinary pupil. (See Chapter X.)

The immediate aim, though not the sole object, of instruction in physics should be to give the power and the habit of using physical knowledge. It should, therefore, on the side of illustrations and applications, be suggestive and directive rather than exhaustive. The pupil should be encouraged to see and think about physical phenomena and physical devices which are outside the classroom; but the teacher should not be expected to bring all such things to his attention and *make* him understand them.

In the way of practical applications of theory, as well as of theory itself, most general text-books of physics contain more, and should contain more, than the ordinary class can be expected to master while in school. The teacher should not be afraid to use his own best judgment, and omit what he feels to be impracticable or comparatively useless. I never find a general text-book of which I can require a class to take every page. The fact is, of course, that the author has in mind a greater variety of readers than is found in any one class, and it is generally easier for the teacher to skip an unnecessary page than to supply a missing one.

Care for Form.

Although in this chapter I have expressly taken it for granted that the teacher will see the need of thorough preparation for what he is to say and what he is to do in the lecture-room, it may be worth while to remark that such preparation will include not merely the parts which are to make up the teacher's performance, but the performance as a whole. The teacher must consider not only what to say and what to do, but when, in what order, each thing is to come. He must think, too, not only of logical sequence, but also of the state of mind and body of his class, following no single line of thought too long, presenting no especially difficult matter when the class is tired.

Physics is, at the best, hard for most minds, young or older; and if the teacher is blessed with the gift, or can by pains acquire the power, of presenting his subject in an attractive way, of making his teaching artistic in form as well as sound in substance, he will win not only the respect of his pupils but, what is perhaps to both sides more stimulating, their admiration.

CHAPTER IX

PHYSICS IN PRIMARY AND GRAMMAR SCHOOLS

REFERENCES.

Bailey, J. H. Inductive Physical Science. Boston, D. C. Heath & Co. 1896. Pp. 105. Qualitative work.

Cooley, L. R. C. Easy Experiments in Physical Science. New York, American Book Co. Pp. 85. Qualitative work.

Gifford, J. B. Elementary Lessons in Physics. Boston, Thompson, Brown & Co. 1894. Pp. 161. Largely qualitative.

Gregory, R. A. and Simmons, A. T. Elementary Physics and Chemistry, First Stage. 1899. Pp. 150. Elementary Physics and Chemistry, Second Stage. 1900. Pp. 140. London and New York, Macmillan.

Harrington, C. L. Physics for Grammar Schools. American Book Co. 1897. Pp. 123. Largely qualitative.

Jackman, W. S. Nature Study for Grammar Grades. New York, Macmillan. 1898. Pp. 407. A book of suggested experiments (for teacher or pupil) and questions.

Loewy, B. A Graduated Course of Natural Science. Parts I. and II. London and New York, Macmillan & Co.

"Nature Study."

I AM quite in sympathy with the not uncommon practice of giving a little, a very little, physics of a descriptive and illustrative kind to young children as a part of what is often called "nature study." But as I have never taught physics to such children in any systematic way, and am not even widely read in the literature of "object lessons," it becomes me to speak with caution in regard to such teaching. I shall venture the suggestion, however, that some of the books in which these lessons are set forth make too little appeal to the experience and the imagination of the pupils.

For example, from an English book which has much to be commended of instruction in regard to common things, I take the following passages, which certainly explain themselves :

"Have this stone, this block of wood, and this piece of iron any shape of their own? Yes, they have, and we cannot alter the shape of either of them. Let one of the boys take the block of wood in his hand, and another the piece of iron, and try to squeeze them into any other shape. He cannot alter the shape of either the wood or the iron with all his pressing."

"Let another boy try with the stone."

"Now put the stone, or the wood, or the iron into a basin, a tumbler, or some such vessel, and let the class see for themselves that these substances do not take the shape of the vessel in which they are placed."

Child's Experimental Knowledge.

I find it difficult to believe that children old enough to go to school would need to make or to see any of the experiments here described, in order to reach the desired conclusion; and it is bad practice to ignore the vast amount of knowledge which comes to every child by mere virtue of his living and having his five senses,— knowledge which becomes a part of him, is blended with his natal instincts, long before he can read and write.

Difficulty with Words.

No small part of the difficulty which young pupils often meet in the study of physics is difficulty with words, due, often, to lack of simplicity or lack of precision in the language of the book or the teacher. For example, what pupil is ever confused as to the fact of "impenetrability," and what pupil is not confused by the word? I was once asked by a grammar-school teacher what I thought of a certain text-book of physics which he was using. I replied that it seemed to me to dwell too much on words and definitions. But the teacher said that it was necessary to make a good deal of effort to get down to the comprehension of his pupils. For example, he had lately spent many minutes in trying to get his class to understand the book definition of uniform velocity, which ran somewhat as follows: *Uniform velocity is such a rate of motion that equal distances are traversed in equal successive intervals of time.* I then suggested that the *idea* of uniform velocity could be given with perfect clearness in a very short time by means of an illustration, and that the analysis and mastery of such a definition as he had been using, though it might be highly profitable to the pupils as an exercise in language, was not the study of physics. This the teacher admitted, but he held that such language lessons are a legitimate use of the time assigned to physics in a grammar school.

This teacher's view may be right, but let us, at least, locate the difficulty properly, and not condemn the study of physics as too hard for grammar schools, merely because many of its simpler truths are often, unnecessarily, expressed too abstrusely. Why could we not say, *Uniform velocity is velocity that is unchanging, neither growing greater nor growing less?*

The habit of overlaboured expression, on which I have been commenting, is often the result of a commendable desire to escape a still worse fault, the habit of indefiniteness, lack of precision of speech and meaning. I have known a class to spend nearly an hour on an elementary exercise, leading up to the hydrostatic press, without any general understanding as to whether the word size, as used with regard to the tubes employed in the experiment, meant diameter or area of cross-section.

Lack of Precision.

An interesting and important question is, whether the study of physics by young pupils should be mainly qualitative or mainly quantitative, that is, whether it should be devoted mainly to the development and illustration of important phenomena, or mainly to the study of numerical laws relating to such phenomena. In my opinion, the little physics taught in primary schools or in the lower grades of grammar schools, should be mostly or wholly of the lecture-table sort, and qualitative. But as soon as the formal study is begun, with laboratory work by the pupils, I am clear that the work should be, I had almost said must be, chiefly quantitative. It is so difficult to design a course of laboratory experiments which will lead the pupil to discover or observe, in any general way, phenomena not previously known to him, so difficult, therefore, to prevent

Qualitative or Quantitative?

qualitative laboratory work from becoming a farce and a bore, in which the wearied teacher points out to each pupil the thing which the latter is supposed to discover, that I have long considered the undertaking unprofitable. Of course it is easy to write out a long list of questions, most excellent if the pupil could find the answers to them, leaving it for the teacher to make the apparatus and devise all details; but how much is really accomplished by such imposing suggestions?

It is true that writers of great ingenuity have undertaken to lay out practical courses of qualitative work, and that some teachers of great zeal are following more or less closely the courses which they have described; but I get from their work, so far as I am familiar with it, at times the impression of tremendous "inductive" feats, surpassing the intuitions of Newton, and at others the impression of an effort rather to occupy the pupil as long as possible with certain simple pieces of apparatus than to teach him as much as possible in a given time. The latter practice reminds one of a box of puzzle blocks with a chart of the figures which, with sufficient ingenuity and time, can be constructed from them. Puzzle blocks are very good indeed in their way, and I am far from asserting that the kind of laboratory work which I have compared to their use is profitless. It does, no doubt, give some manipulative skill, and it gives some practice in keeping a record of observations, but, as a means of getting forward with the study of physics, I believe that it can be greatly improved upon.

Too Slow Progress.

The practice of dwelling unnecessarily long on things familiar and essentially simple, of discussing them at great length and laboriously writing out observations upon them, is a vicious one; for it inculcates a habit of pottering, and is quite as likely to result in confusion of ideas as in lucidity. The fact that a pupil cannot give a clear account of some particular fact or law is no sure proof that he has not spent too much time on it. It is possible to gaze at one's own name until it looks unfamiliar and weird.

Movement, a certain sense of progress, is essential to the

best working of the pupil's mind, which, like a bicycle, simply lies down if it is kept too long in one spot. It is better to maintain this progress, even with the certainty that some things will be passed by unseen, and that many of the things seen will be forgotten, than to lose headway and the alertness which goes with it. Many repetitions are necessary for the mastery of certain truths; but these repetitions should not all come at one stretch. An occasional brief return to the difficult point, when the mind is fresh, is better in many cases than the attempt to level every obstacle and clear up every doubt at the first progress.

Quantitative Laboratory Work in Grammar Schools.

For eight or nine years now the grammar schools of Cambridge, Massachusetts, have maintained a course of quantitative laboratory work for pupils of the ninth grade, averaging perhaps fourteen years of age. The titles of the exercises in this course are, for the most part, such as are to be found in the First Part of the Harvard *Descriptive List* or the corresponding list of the National Educational Association (see Chapter X.); and the method of performance of these exercises follows pretty closely the directions given in the *Descriptive List*. The pupils, however, do not have these or any printed directions before them in doing their work, nor have they, in fact, any text-book of physics.

Exercises Taken.

Apparently the teachers prefer not to have a book in the hands of the pupils. The time allowed for the whole course is only two school periods, of 40 minutes each, a week for one school year, and physics is treated as one of the minor studies of the grammar school course. Under these conditions the teachers apparently feel that it is hardly worth while to take up a text-book, some parts of which might be too difficult or too laborious for their pupils. I think, too, that they find a certain legitimate satisfaction in lecturing to their classes in this study, which is more objective than most others with which they have to do.

No Text-book.

It must be admitted that when these Cambridge pupils, after dropping the study of physics entirely for two years, resume it in the third year of the high school course, there is generally not very much immediately visible in their minds as the result of their previous work in this subject. But in what study will the direct product of two school periods per week for one grammar school year show to great advantage two years later? Arithmetic, geography, history? In no study that can be named, unless circumstances are such as to keep the lessons of that study frequently in practice.

Permanence of Results?

Many of the things learned by a boy in such a course of physics as that indicated above, will go into frequent practice in his every-day life. He will, therefore, probably remember his physics as well as he remembers anything on which so little time has been spent. It will be best, if not necessary, to go all over the same ground again in the high school; but that is no proof whatever that the first study has not been profitable. Such preliminary study is like the coat of oil which is laid on wood to prepare it for varnishing. The oil dries in and disappears, but the varnish, the show coat, sticks because the oil has gone before it.

But the question remains, whether such work had better be done by a grammar school class. There is a chance for misunderstanding here. The question is not, whether the same list of experiments done later will teach more, for it must be granted that almost any study pursued at the grammar school age would yield larger results with an equal expenditure of time two or three years later. The question is, whether this course, or some such course, of physics is more profitable to the class as a whole than anything which could, or would, take its place.

I am, possibly, too much influenced by the circumstances of the case to give this question a judicial answer. If all the pupils were to go forward into a higher school, I should perhaps answer it in the negative; but a comparatively small proportion of them do this. Is it wise, is it fair, to let the great majority of public-school children close their school life without any formal study of natural science? Are school authorities sufficiently sure, for example, of the superior profitableness of the back part of the arithmetic, partial payments, etc., to warrant them in preserving all its commercial features to the exclusion of natural science?

The following paragraphs are written by Mr. Frederick S. Cutter, the master of the Peabody Grammar School of Cambridge:

"The time allotted to the subject [of physics] is one hour and twenty minutes a week throughout the school year, of which thirty minutes is for laboratory work and fifty minutes for discussion and lecture-table instruction in the classroom. The class is divided [for laboratory work] into sections of sixteen (or less) pupils each."

Comments by School-Master.

"Time for the introduction of physics, and also geometry, was obtained in the revision of the course of study by completing the subject of geography in the eighth grade and by modifying the requirements in arithmetic. The one hour and twenty minutes a week devoted to physics is supplemented by a part of the time assigned to language work, when written compositions are prepared by the pupils in which accounts of their experiments are given from the notes taken in the laboratory. These compositions are usually illustrated, for children as a rule like to write about what they have performed, and take pleasure in the adornment of their papers. Thus the subjects of physics, language, and drawing are most profitably correlated.

"Before the introduction of laboratory physics there were some who feared that a serious difficulty would be the time and labour required of the teacher in preparation for an experiment to be performed by sixteen children, and afterwards in putting the things away. But in practice it is found that the teacher can be largely relieved by several of the most trustworthy pupils, who are always glad to offer their services as assistants. To one can be given entire charge of the sixteen large glass jars, the filling, the emptying, and the putting away in proper condition; to another can be given the care of the sixteen overflow cans; to another the care of the spring-balances, etc. The children selected will profit by the responsibility they assume, and will take increased interest in the work, their influence being favourably felt throughout the class. If the teacher announces in the morning session what will be needed for the experiment in the afternoon, the pupils can get everything in readiness during the noon intermission."

CHAPTER X

PHYSICS IN VARIOUS KINDS OF SECONDARY SCHOOLS

REFERENCES.

Laboratory Manuals:

Manuals included in text-books and not published separately are not here named.

Adams, C. F. Physical Laboratory Manual for Secondary Schools. Chicago & New York, Werner School Book Co. 1896. Pp. 183.

Allen, C. R. Laboratory Exercises in Elementary Physics. New York, Henry Holt & Co. 1892. Pp. 277.

Ames, J. S. and Bliss, W. J. A. Manual of Experiments in Physics. American Book Co. 1898. Pp. 544.

Ayres, F. H. Laboratory Exercises in Elementary Physics. London, E. Arnold. New York, D. Appleton & Co. 1901. Pp. 193. To accompany Henderson and Woodhull's Elements of Physics.

Chute, H. N. Physical Laboratory Manual. London, Isbister & Co. Boston, D. C. Heath & Co. 1894. Pp. 218.

Crew, H. and Tatnall. Laboratory Manual of Physics. London and New York, Macmillan. 1902. Pp. 234.

Gage, A. P. Physical Experiments. Boston, Ginn & Co. 1897. Pp. 195.

Gage, A. P. Physical Laboratory Manual and Note-book. 1890. Pp. 244. Boston, Ginn & Co.

Glazebrook, R. T. and Shaw, W. N. Practical Physics. London and New York, Longmans, Green & Co. New Edition, 1901. Pp. 659. Too difficult for school use.

Henderson, John. Elementary Physics. London and New York, Longmans, Green & Co. 1895. Pp. 132. For colleges rather than schools.

Henderson, C. H. and Woodhull, J. F. Physical Experiments. New York, Appleton & Co. 1900. Pp. 112. To accompany the Authors' Elements of Physics and sometimes bound with the Elements.

Hopkins, W. J. Preparatory Physics. Longmans, Green & Co. 1894. Pp. 147.

Hortvet, J. A Manual of Elementary Practical Physics for High Schools. Minneapolis, H. W. Wilson.

Kelsey, W. R. Physical Determinations. London, Arnold. New York, Longmans, Green & Co. 1901. Pp. 328.

Nichols, Smith and Turton. Manual of Experimental Physics. Boston, Ginn & Co. 1899. Pp. 324.

Rintoul, D. An Introduction to Practical Physics. London and New York, Macmillan. 1898. Pp. 166. Mensuration, Mechanics, and Heat.

Schuster, A. and Lees, C. H. Intermediate Course in Practical Physics. London and New York, Macmillan & Co. 1896. Pp. 248. Describes a course given in Owens College.

Stewart, B. and Gee, W. W. H. Elementary Practical Physics. Vol. I. General Processes. 1893. Pp. 295. Vol. II. Electricity and Magnetism. 1896. Pp. 503. Vol. III. Practical Acoustics (by C. L. Barnes). 1897. Pp. 214. London and New York, Macmillan & Co. Rather beyond school class use.

Stratton. Outline of a Course in General Physics for Secondary Schools: SCHOOL REVIEW. January, 1898. 7 pages.

Watson, Wm. Elementary Practical Physics. A Laboratory Manual for Beginners. London and New York, Longmans, Green & Co. 1896. Pp. 238.

Woodhull, J. F. First Course in Science. Vol. I. Book of Experiments. Henry Holt & Co. 1893. Pp. 79. Light only. Vol. II. is a text-book to accompany the manual. Pp. 133.

Worthington, A. M. Physical Laboratory Practice. London, Longmans, Green & Co.

General Text-books, Mainly for Teachers' Use:

Anthony, W. A. and Bracket, C. F. Elementary Text-book of Physics. Revised by Magie. New York, John Wiley & Sons. 1897. Pp. 512.

Barker, G. F. Physics. Advanced Course. London, Macmillan & Co. New York, Henry Holt & Co. 1892. Pp. 902.

Daguin, P. A. Traité Élémentaire de Physique. Paris. 3 vols. Contains much descriptive and historical matter.

Daniell, A. A Text-book of Physics. London and New York, Macmillan & Co. 1895. Pp. 782.

Deschanel, A. P. Natural Philosophy. Translated and revised by Everett. London, Blackie & Son. New York, D. Appleton & Co. 4 vols. The 3d vol., Electricity and Magnetism, issued in 1901.

Ganot's Physics. Translated and edited by E. Atkinson. London, Longmans, Green & Co. New York. Wm. Wood & Co. Pp. 1115.

Ganot's Popular Natural Philosophy. Translated and edited by E. Atkinson. Longmans, Green & Co. 1900. Pp. 752.

Hastings, C. S. and Beach, F. E. Text-book of General Physics. Boston, Ginn & Co. Pp. 768.

Lehfeldt, R. A. A Text-book of Physics, with Sections on the Applications of Physics to Physiology and Medicine. London, Arnold. New York, Longmans, Green & Co. 1902. Pp. 312.

Nichols, E. L. and Franklin, W. S. Elements of Physics. Vol. I. Mechanics and Heat. 1898. Pp. 219. Vol. II. Electricity and Magnetism. 1896. Pp. 272. Vol. III. Light and Sound. 1899. Pp. 201. London and New York, Macmillan.

Watson, Wm. A Text-book of Physics. London and New York, Longmans, Green & Co. 1899. Pp. 896.

Special Treatises, Mainly for Teachers' Use:

Abney, W. D. W. Treatise on Photography (Text-books of Science). London and New York, Longmans, Green & Co. 1901. Pp. 425.

Anderson, J. Strength of Materials and Structures (Text-books of Science). London and New York. Longmans, Green & Co. 1892. Pp. 307.

Benjamin, P. A History of Electricity. New York, John Wiley & Sons. Pp. 611.

Carpenter, R. C. Heating and Ventilation of Buildings. New York, John Wiley & Sons.

Goodeve, T. M. Text-book on The Steam Engine and on Gas Engines. London, C. Lockwood & Sons. New York, D. Van Nostrand Co.

Halliday, G. Steam Boilers. London, Edward Arnold. New York, Longmans, Green & Co. 1897. Pp. 392.

Hardin, W. L. Liquefaction of Gases. London and New York, Macmillan. 1899. Pp. 250.

Harrington, M. W. About the Weather. (Home Reading Books.) New York, D. Appleton & Co. 1901. Pp. 246.

Hawkins, C. C. and Wallis, F. The Dynamo. London, Whittaker & Co. New York, Macmillan & Co. 1896. Pp. 526.

Holmes, G. C. V. The Steam Engine. (Text-books of Science.) London and New York, Longmans, Green & Co. 1897. Pp. 528.

Holden, E. S. Stories of the Great Astronomers. 1901. Pp. 255. The Family of the Sun. 1899. Pp. 252. (Home Reading Books.) New York, D. Appleton & Co.

Hopkins, W. J. Telephone Lines and Their Properties. 1886. Pp. 272. The Telephone. 1898. Pp. 83. New York and London, Longmans, Green & Co.

Jackson, D. C. and J. P. Alternating Current and Alternating Current Machinery. London and New York, Macmillan. 1902. Pp. 482. "For Artisans, Apprentices, and Home Readers."

Jones, H. C. Theory of Electrolytic Dissociation. London and New York, Macmillan. 1900. Pp. 289.

Le Blanc, M. Electrochemistry. London and New York, Macmillan. 1896. Pp. 284.

Maxwell, J. C. Theory of Heat. With corrections and additions by Lord Rayleigh. London and New York, Longmans, Green & Co. Pp. 348.

Preece, W. H. and Sivewright, J. Telegraphy. London and New York, Longmans, Green & Co. 1897. Pp. 417.

Preston, T. The Theory of Heat. 1894. Pp. 719. The Theory of Light. 1901. Pp. 586. London and New York, Macmillan.

Shelley, C. P. B. Work-Shop Appliances. London and New York, Longmans, Green & Co. 1897. Pp. 377.

Shenstone, W. A. The Methods of Glass-Blowing. London and New York, Longmans, Green & Co. New ed., 1902. Pp. 86.

Slingo, W. and Brooker, A. Electrical Engineering for Electric Light Artisans and Students. London and New York, Longmans, Green & Co. 1898. Pp. 780.

Stretton, C. E. The Locomotive and Its Development. London, C. Lockwood & Sons.

Thompson, S. P. Elementary Lessons in Electricity and Magnetism. 1894. Pp. 638. Light Visible and Invisible. 1897. Pp. 294. London and New York, Macmillan.

Treadwell, A. The Storage Battery. London, Whittaker & Co. New York, Macmillan & Co. 1898. Pp. 257.

Whetham, W. C. D. Solution and Electrolysis. Cambridge, University Press. New York, Macmillan. 1895. Pp. 296.

Wilson, E. Electrical Traction. London, Edward Arnold. New York, Longmans, Green & Co. 1897. Pp. 253.

Wright, M. R. Sound, Light, and Heat. Pp. 272. Heat. Pp. 346. London and New York, Longmans, Green & Co.

Yorke, J. P. Magnetism and Electricity. London, Edward Arnold. New York, Longmans, Green & Co. 1899 Pp. 264.

ON the subject of this chapter we have something approaching the authority of official utterance in the various publications made by the National Educational Association during the past ten or twelve years.

College Entrance Physics of the National Educational Association.

General Recommendations.

The general definition or description of the type of physics course, preparatory for college, which is approved by the National Educational Association, is shown by the following extract from the report of its committee on College Entrance Requirements, which report was published in 1899 by the authority of the association.

"Your committee suggests that an effective working basis for a secondary school course in physics would be attained by planning such a course substantially in accordance with the following propositions:

"1. That in public high schools and schools preparatory for college physics be taught in a course occupying not less than one year of daily exercises, more than this amount of time to be taken for the work if it is begun earlier than the next to the last year of the school course.

"2. That this course of physics include a large amount of laboratory work, mainly quantitative, done by the pupils under the careful direction of a competent instructor and recorded by the pupil in a note-book.

"3. That such laboratory work, including the keeping of a note-book and the working out of results from laboratory observations, occupy approximately one-half of the whole time given to physics by the pupil.

"4. That the course also include instruction by text-book and lecture, with qualitative experiments by the instructor, elucidating and enforcing the laboratory work, or dealing with matters not touched upon in that work, to the end that the pupil may gain not merely empirical knowledge, but, so far as this may be practicable, a comprehensive and connected view of the most important facts and laws in elementary physics.

"5. That college-admission requirements be so framed that a pupil who has successfully followed out such a course of physics as that here described may offer it toward satisfying such requirements." [1]

History of the N.E.A. Report.

The report from which the preceding extract is taken was approved in the year following its publication, in the following resolution: "*Resolved*, That the Departments of Secondary and Higher Education of the National Educational Association commend the report of the special Committee on College-Entrance Requirements as offering a basis for the practical solution of the problems of college admission, and recommend the report to the colleges of the country."

The report under discussion consisted of two parts, a shorter part, for which the committee itself took the responsibility, and

[1] These five propositions are substantially a repetition of recommendations made to the General Committee by the Committee on Physics, the membership of which is given later.

from which the extract above given is taken, and a longer part containing many details offered in the reports of various sub-committees. In this supplementary part of the report the matter relating to physics, quoted in full below, is little more than the Table of Contents of the Harvard *Descriptive List* (see Chapter V.) and two paragraphs taken, almost without change, from the Introduction to that list. As this fact is capable of misinterpretation by those not fully acquainted with the circumstances, I shall give some account of the matter here.

As chairman of the committee mentioned in the quotation following, I had reported to Dr. Nightingale, the chairman of the general Committee on College Entrance Requirements, that, inasmuch as the Physics Committee was made up largely of gentlemen who had written text-books — very different text-books — for schools, or were in the way of writing such books, I could not undertake to get from them any general agreement as to the details of what a course in schools should be. But, as Dr. Nightingale insisted on some kind of a report on this subject, I at last sent him what is practically a description of the Harvard requirement in laboratory work, as my individual report, at the same time notifying the other members of the committee of what I had done, and requesting each of them to take corresponding individual action, if he felt moved to do so. The quotations which follow show the result.

Personnel of N. E. A. Physics Committee.

"The Committee on Physics of the Science Department of the National Educational Association did not submit a regular report signed by the members of the committee. These were: Professor E. H. Hall, Harvard University, chairman; Professor H. S. Carhart, University of Michigan, Ann Arbor; R. B. Fulton, Chancellor, University of Mississippi; C. L. Harrington, Sachs' Collegiate Institute, New York, N. Y.; Julius Hortvet, East Side High School, Minneapolis, Minn.; C. J. Ling, Manual Training School, Denver, Colo.; Professor E. L. Nichols, Cornell University, Ithaca, N. Y.; E. D. Pierce, Hotchkiss

School, Lakeville, Conn.; Professor Fernando Sanford, Leland Stanford, Jr., University, Cal.; Professor B. F. Thomas, Ohio State University, Columbus; Edward R. Robbins, Lawrenceville School, Lawrenceville, N. J.

"The basis of a report, suggested by Professor Hall, and consisting of a list of laboratory experiments, is given below. Comments by the members of the committee, in case they dissented from any part of this, were to be sent at once to the chairman of the Committee on College-Entrance Requirements. It may be assumed that the list met with the approval of those who did not so indicate dissent. Such comments as have been received are given after Professor Hall's statement.

"**Outline of Laboratory Work in Physics for Secondary Schools.**

"At least thirty-five exercises, selected from a list of sixty or more, not very different from the list given below. In this list the divisions are mechanics (including hydrostatics), light, heat, sound, and electricity (with magnetism). At least ten of the exercises selected should be in mechanics. The exercises in sound may be omitted altogether; but each of the three remaining divisions should be represented by at least three exercises.

Detailed List of Exercises.

"The division of the list into a first part and a second part is intended to facilitate and encourage beginning the study of physics very early in the school course. Most of the exercises in the first part have proved to be within the power of boys of fourteen or fifteen years, although older pupils can do them more readily, as they can do all other work except tasks of pure memory. The cost of apparatus for the exercises of the first part is very small.

"**First Part.**

PRELIMINARY EXERCISES.

[Recommended, but not to be counted.]

A. Measurement of a straight line.

B. Lines of the right triangle and the circle.

C. Area of an oblique parallelogram.
D. Volume of a rectangular body by displacement of water.

Mechanics and Hydrostatics.

1. Weight of unit volume of a substance.
2. Lifting effect of water upon a body entirely immersed in it.
3. Specific gravity of a solid body that will sink in water.
4. Specific gravity of a block of wood by use of a sinker.
5. Weight of water displaced by a floating body.
6. Specific gravity by flotation method.
7. Specific gravity of a liquid: two methods.
8. The straight lever: first class.
9. Centre of gravity and weight of a lever.
10. Levers of the second and third classes.
11. Force exerted at the fulcrum of a lever.
12. Errors of a spring balance.
13. Parallelogram of forces.
14. Friction between solid bodies (on a level).
15. Coefficient of friction (by sliding on incline).

Light.

16. Use of Rumford photometer.
17. Images in a plane mirror.
18. Images formed by a convex cylindrical mirror.
19. Images formed by a concave cylindrical mirror.
20. Index of refraction of glass.
21. Index of refraction of water.
22. Focal length of a converging lens.
23. Conjugate foci of a lens.
24. Shape and size of a real image formed by a lens.
25. Virtual image formed by a lens.

Second Part.

Mechanics.

26. Breaking strength of a wire.
27. Comparison of wires in breaking tests.

28. Elasticity: stretching.
29. Elasticity: bending; effects of varying loads.
30. Elasticity: bending; effects of varying dimensions.
31. Elasticity: twisting.
32. Specific gravity of a liquid by balancing columns.
33. Compressibility of air: Boyle's law.
34. Density of air.
35. Four forces at right angles in one plane.
36. Comparison of masses by acceleration test.
37. Action and reaction: elastic collision.
38. Elastic collision continued: inelastic collision.

Heat.

39. Testing a mercury thermometer.
40. Linear expansion of a solid.
41. Increase of pressure of a gas heated at constant volume.
42. Increase of volume of a gas heated at constant pressure.
43. Specific heat of a solid.
44. Latent heat of melting.
45. Determination of the dew-point.
46. Latent heat of vaporization.

Sound.

47. Velocity of sound in open air.
48. Wave-length of sound.
49. Number of vibrations of a tuning fork.

Electricity and Magnetism.

50. Lines of force near a bar magnet.
51. Study of a single-fluid galvanic cell.
52. Study of a two-fluid galvanic cell.
53. Lines of force about a galvanoscope.
54. Resistance of wires by substitution: various lengths.

55. Resistance of wires by substitution: cross-section and multiple arc.
56. Resistance by Wheatstone's bridge: specific resistance of copper.
57. Temperature coefficient of resistance in copper.
58. Battery resistance.
59. Putting together the parts of a telegraph key and sounder.
60. Putting together the parts of a small motor.
61. Putting together the parts of a small dynamo.

"Professor Carhart suggests forty experiments similar to these. Twenty-four of these coincide exactly in title with items in the above list. The following fourteen are new, but many of them are probably implied in the list of sixty-one:

The Jolly balance.
Laws of the pendulum.
Pressure.
Curve of magnetization.
Action of current on needle.
Fall of potential in conductor.
E. M. F. of cell.
The tangent galvanometer.
Velocity of sound in solids (Kundt).
Law of length for strings (sound).
Law of diameter for strings (sound).
Law of tension for strings (sound).
Law of reflection (light).
Measurement of angle of prism (light)."[1]

Action of Middle States Board.

The next following quotation is an extract from the *Definition of Requirements* issued by the College Entrance Examination Board of the Middle States and Maryland, February 1, 1901.

[1] In the NEW ENGLAND JOURNAL OF EDUCATION for December 26, 1901, and January 2 and 9, 1902, Mr. Stratton D. Brooks, High School Visitor for the University of Illinois, has, under the title, "Suggested List of Experiments in Physics," worked over this N. E. A. list and given corresponding references to several well-known text-books.

"8 Physics.

"The requirement in physics is based on the report of the Committee on Physics of the Science Department of the National Educational Association.

"It is recommended that the candidate's preparation in physics should include

"*a.* Individual laboratory work, comprising at least thirty-five exercises selected from a list of sixty or more, not very different from the list given below.

"*b.* Instruction by lecture-table demonstrations, to be used mainly as a basis for questioning upon the general principles involved in the pupil's laboratory investigations.

"*c.* The study of at least one standard text-book, supplemented by the use of many and varied numerical problems, 'to the end that the pupil may gain a comprehensive and connected view of the most important facts and laws in elementary physics.'"

The list of titles of experiments which follows this passage in the original context is precisely the same as that numbered from 1 to 61 in the Report of the National Education Association and in the Harvard *Descriptive List.*

Prevalence of Such a Course?

It appears, then, that we have, in the course of work outlined by the preceding quotations, a type of college entrance requirement in physics which is tolerably well defined and widely approved. Whether this type is established and maintained as generally as it is approved, may be an open question. In that part of the country which comes under my personal observation, it is very generally established. But in this same region the boys who go through a high school course, without having preparation for college in view, do not, as a rule, take just this course of physics. They take one which is more "practical" or more "general" or more "popular," almost always, I believe, a course that involves less close attention and hard thinking. This fact naturally raises a number of questions. Is the college requirement, as interpreted and maintained by Harvard, for example, more severe than it should be? Are its applications to every-day life too remote? Does it require too much use of mathematics? Does it have too large a proportion of painstaking laboratory work, and too little in the way of lecture-room exhibitions? In particular, should the course make great use of the projecting lantern, with a large collection of interesting "slides," illustrating scientific objects of general or local importance? Do the teachers who devise the courses of physics study followed in "English high schools," and other schools of the same general character, virtually express an unfavourable judgment of the college requirement physics for boys who are not to go to college?

Essential Difference in Schools.

It is possible that some of these questions would be and should be generally answered in the affirmative, but this is not the inevitable conclusion. There is still the possibility that those who have advocated[1] the same work for boys who are to go to college as for boys who are not to go to college have overlooked one very important fact, namely, that the two sets of boys may not be just alike in their mental traits and attainments. As a rule, so far as my observation and inquiry have gone, they are notably different, the boys who naturally go to an "English high school" being less scholarly and more narrowly utilitarian in their views than their contemporaries and associates who naturally go to a "Latin school."

Moreover, I can see little prospect of the disappearance or even the diminution of this unlikeness. The now well established practice of teaching, in the English high schools, such arts as book-keeping, short-hand, and typewriting, inevitably draws into these schools a numerous class of boys and girls who by birth and home influence have received little of scholarly capacity or impulse. Yet their parents demand, and with good show of reason, that if public money is spent to advance the few

[1] See the Report of the Committee of Ten, which is very emphatic on this point.

to the doors of college, with the comparatively profitable learned professions in view beyond, public money shall be spent to advance the many toward the practice of their useful and honourable, if less distinguished, vocations. The ordinary city high school will therefore continue to have a general class of pupils who are not capable of going side by side with the pupils of the Latin schools, — a class who have left the grammar schools comparatively old, and will leave the high school at a lower intellectual level than their Latin school contemporaries, unless the course of the former school is made longer than that of the latter, which is not likely to be the case.

Stimulus of College Requirements.

Furthermore, even if the natural difference in kind of pupils did not exist, the fact that the pupils in one school are preparing to meet requirements set by an authority outside the school, while the pupils in the other school are without this stimulus, will probably always keep the general standard of the work higher in the former school than in the latter. It is very doubtful whether local authority, or even the authority of any state board of education, unsupported by the strongly asserted requirements of colleges, can ever be depended on to keep the general standard of graduation from the high school up to the proper level of admission to college. A school which sends but few boys to college will prepare but few boys for college.

What High Schools Should Do.

What, then, should such a school undertake to do in physics? Should it follow the college preparatory course as far as it can, taking half or two-thirds of it, for example; or should it maintain a course designed with especial reference to the character and aims of its own pupils? I cannot doubt that the latter alternative will prevail, and ought to prevail, though it should be the constant effort of all school authorities to broaden and elevate, so far as this may be practicable, the ideas of their pupils as to what is attainable and what is worth while.

Without directly following the college requirement course in physics, the high school course has been profoundly influenced by it, and will doubtless continue to be so. On the other hand, it behooves those of us who have most to do with the college requirement, to keep watch of the development of physics in the high schools, with the hope of finding therein examples which we may profitably follow.

For the high school course in physics, as distinguished from the college requirement course, there is not, so far as I am aware, any general description arrived at by formal consensus of opinion. The following extract from the official description of the physics work in the high school of Brookline, Massachusetts, gives account of a course developed by Mr. John C. Packard, the teacher of physics in that school, which is in marked and interesting contrast with the college requirement course:

Physics in Brookline High School.

"There are two courses in physics.

"1. The so-called Popular Course, the fundamental aim of which is:

"(a) To develop in the pupil the habit of steady, persistent, logical thinking;

"(b) To render him fairly intelligent in reference to his own scientific environment;

"(c) To beget a sense of power in his own ability to appreciate scientific truth and to draw legitimate conclusions from simple data;

"(d) To teach him to apply the elements of Algebra and Geometry to the problems of daily life, and finally

"(e) To arouse within him a deep sense of appreciation of all that modern science has done and is still doing for the comfort and convenience of the race.

"With these ends in view the head of the department in common with many others has discovered that but very little reliance can be placed upon the ordinary text-book, since so few opportunities are given in the average manual for any original independent thinking and since in general such books contain so little of anything like a practical application of the principles of physics to the phenomena of daily life. He has felt obliged therefore to substitute for such text-books, as others have, a special manual, as yet in manuscript form, in which the student is told as little as possible directly, but is given, practically, a series of original exercises in Mechanics, Optics, and Electricity which he is to work out by the aid of a set of simple apparatus, his mathematical instincts, and his own brain, and apply in a continuous sequence suggested by an abundance of questions, problems and references to the affairs of daily life.

"The aim is to be thoroughly practical. In Hydraulics, for instance, more attention is paid to the water-meter, the simple motor, and the turbine than to the lifting pump, the ram and the breast-wheel, as the average man is more likely to see and use the former than the latter series. In Optics again, the camera, the opera glass, and the spyglass are dealt with more fully than the telescope and the compound microscope for the same reason.

"Continual reference is made to the current literature of the day and to the science of Boston and vicinity.

"It is intended that a series of illustrated lectures shall accompany the course giving a brief summary of the history of Physics and a glimpse of the wonderful scientific achievements of our own age.

"The work is distributed somewhat as follows: —

"September, October, November, — Mechanics, including Hydrostatics and Pneumatics.

"December, January, February, — Optics.

"March, April, May, — Electricity.

"June, — Review.

"Toward the close of the school year special topics are suggested for more exhaustive treatment than is possible in the regular classroom work. Each pupil is expected to choose one or more of such topics and to present an illustrated paper upon the subject selected, at the end of the year.

"Among the topics recently suggested may be mentioned the following: —

1. Mechanics of the Clock.

2. Mechanics of the Bicycle.
3. Mechanics of the Sewing Machine.
4. Hughes' Induction Balance.
5. The Microphone.
6. Consumption of Gas, Water and Electricity in the Household.
7. Testing a Water-Meter.
8. The Fire-alarm System of Brookline.
9. School-room Ventilation.
10. The Long-Distance Telephone.
11. The Transformer.
12. The Gas-Engine.
13. The Horse Power of an Electric Motor.

"This entire course, extending over one year's time, is required of the sub-classical, the scientific and the manual training pupils and at least one-half the course, i. e., the first five months, of the classical.

"The time is equally divided between laboratory and lecture-room work, each requiring two periods per week beside the usual preparation for a full study.

"Complete notes are kept by the pupils, of both the laboratory and the lecture work. These notes are inspected from time to time by the instructor," etc.

Problem to be Solved by Experience.

I am far from asserting that the course outlined by Mr. Packard is not better for the average high school pupil, boy or girl, than the college preparatory course, which also is given in the same school. Mr. Packard and others who, like him, have worked out the problem of general high school physics approximately to their own satisfaction on somewhat new lines, will do a service to the public by putting the results of their experimentation into the form of text-books or manuals available for all teachers. These books may or may not prove to be generally acceptable and usable; but in any case they will be an important contribution to that vigorous trying-out process through which all methods of science teaching are now going in the schools of this country.

The word of caution which I would give to those who aim especially to make their teaching "practical" is, that they should beware of encouraging the idea, which many of their pupils are only too much inclined to hold, that the object of schooling is to give a certain final and sufficient store of knowledge and not, rather, so to fit the pupil that he may, after his school days are over, go on increasing in knowledge, finding constantly new uses for that stock of elementary fundamental ideas which a well devised school course should inculcate. To this suggestion the teacher will perhaps reply that the average high school pupil has not sufficient initiative and imagination to find for himself the use of abstract ideas, and that the attempt to implant such ideas in a mind essentially concrete is labour and opportunity lost. I have no confident answer to make to such an assertion. The problem here is to find the right proportion of those constituents which all admit to be necessary. There is no hard and fast rule to be laid down.

CHAPTER XI

ON THE PRESENTATION OF DYNAMICS

REFERENCES.

Magnus, P. Lessons in Elementary Mechanics. London and New York, Longmans, Green & Co. 1892. Pp. 377.

Maxwell, J. C. Theory of Heat. London and New York, Longmans, Green & Co. Chapter IV.

I HAVE not undertaken to give in this book a pedagogic treatment of the various parts of elementary physics; but there is one part, namely, dynamics, so fundamental yet so often neglected or badly taught, that I propose to give it especial attention here.

Difficulty and Importance of the Subject.

It must be admitted once for all that the elementary ideas involved in questions of acceleration are difficult for the ordinary mind to grasp. The formulas, at least for cases of uniform acceleration, are very simple, but the primary conceptions underlying these formulas, the definite notions of force, momentum, and kinetic energy, the ordinary student rarely masters and retains. Should we, therefore, give up the attempt to teach this part of physics in school courses, or the early courses in college, and content ourselves with giving, in mechanics, the statical aspect only?

I fear that many teachers will answer this question in the affirmative, but I am not yet ready to do so. We cannot afford to avoid everything that is difficult for the average boy, or practically impossible for the dull boy. We must conduct our classes with some regard to the most vigorous minds among our pupils; and such minds will find, in the broadening of their vision through the study of dynamics, perhaps the most profitable part of all their training in physics. How can we be content to let a boy of eighteen years or older leave our classrooms without having had an opportunity to learn the meaning of the term energy, as strictly used, — a conception without which all endeavour to understand and use the law of conservation of energy, the grandest, yet one of the simplest, of the generalizations of physical science, is feeble, if not futile?

But may not a man be useful and happy who does not understand the conservation of energy? Yes, if he knows that he does not understand it, and does not profess to understand it. But this law is peculiarly one which many people talk about, and fancy themselves to understand, while their whole notion of it is so vague that it is quite as likely to lead them wrong as right when they would make any application of it. The law lies all about us, and nearly every one has some not altogether false idea of its meaning, some fairly good illustration of it at command; but understand it he certainly does not, if he has not mastered the meaning of certain little words, and certain short formulas, the full significance of which is not made plain by the experience and conversation of every-day life. Great thinkers groped long for the full meaning of the law, discoursing meanwhile of the "conservation of force," and using "force" sometimes in its proper sense, sometimes in the sense of "energy," feeling their own confusion of ideas, but unable to see just where their trouble lay.

Difficulty Increased by Poor Teaching.

I began this chapter by admitting the difficult nature of the ideas used in dynamics. I believe, however, that the effort of mastering these ideas will be less for the next generation than for the present, not through any considerable growth in the power of the human brain within a few decades, but because good methods of instruction, if we can establish and maintain them, will gradually produce teachers thoroughly competent to guide their pupils through the initial difficulties of dynamics. That all teachers of physics are not yet in this condition, a very brief tour of visits to classrooms will show.

Not long ago, in a flourishing city school, I heard part of a recitation on the meaning and application of the law of acceleration, $f = \frac{m \times v}{t}$, where v is the velocity imparted in t seconds by the force f to the mass m. The teacher remarked to me when I entered the room that he found it hard to get his pupils to understand the *dyne*. I expressed sympathy; but, in the discussion with pupils which presently followed, the teacher repeatedly gave his approval to the following statement: *A dyne is the force which will move one gram one centimeter in one second.*

Instances.

In another school, — an excellent school, — I heard a teacher, after giving his pupils to understand that sliding friction is somewhat less with high velocity than with low velocity, — a very doubtful proposition in itself, — explain this alleged fact by declaring that the momentum of the moving body helps to carry it over the frictional obstacles. The experiments under discussion were such as involved uniform velocity of the sliding body.

In still another excellent school I heard a teacher discuss the pressure in a siphon, in operation, as if the question were one of simple hydrostatics, assuming the pressure at a given level, in the stream within the siphon, to be just as great as the pressure in the still water at the same level outside the siphon, thus neglecting altogether the difference of pressure used in giving momentum to the water entering the tube.[1]

The law that action is equal to reaction and opposite in direction, is so very simple in form and so easily remembered verbally, that probably most people who have ever heard it

[1] The case of pressure in the siphon, during flow, seems rather too difficult for profitable discussion, in detail, with a school class or a young class in college. I think it better to keep to the static aspect, consider the siphon filled, with its lower end closed for the moment, and merely show that the pressure within the siphon at this end at this moment is greater than the atmospheric pressure, so that water must flow out as soon as the tube is opened. Yet some clear elementary conception of the dynamics of the flowing water is needed, in order to enable the teacher of young pupils to see why he had better leave that matter untouched by his class.

think they understand it. Yet there is plenty of evidence that teachers sometimes fail to realize and apply it, even in simple cases of collision of bodies. There is a certain experiment or set of experiments, to the devising of which I have given a great deal of care and thought, intended to illustrate the fact that the algebraic sum of the momenta of two bodies is the same after their collision as before, and that this rule holds true as well for ivory balls with putty interposed as for ivory balls in naked shock. Yet once a teacher of considerable experience, who now holds and deserves an important position in the school system of a large city, complained to me that this set of experiments was a comparative failure, because, according to his observations, it seemed to indicate that inelastic bodies preserved their total momentum as well as elastic bodies. I explained the situation to him in a word; he thanked me heartily, and has, I feel sure, ever since found that particular experiment easier to deal with and more profitable to discuss than it used to be.

I found, too, that another teacher, a well-known man, observing that the total momentum after collision was usually, by the somewhat defective method of estimation prescribed in the experiment referred to, made to appear slightly less than the total momentum before collision, was in the habit of teaching his pupils that the difference found was due to the loss of momentum (or energy?) in the production of heat at the collision. Of course, a teacher in such confusion of mind, as to the relations of momentum and energy, would make a muddle in the minds of his pupils.

Lest these instances of faulty teaching should be considered invidious, let me say that I have long since come to the conclusion that it is unfair and unsafe to condemn any teacher for any single mistake, however glaring.

It is plain that a considerable part of our trouble with elementary dynamical notions comes from our unfortunate, but at present unavoidable, multiplicity of force-units. We have, at the least, the pound and the gram as gravitation units of force, the same names being used also for units of mass, and the poundal and the dyne as absolute, or acceleration, units of force. Some generations hence the pound and the poundal may have disappeared from common use, the decimal system of weights and measures being then fully established; but it is doubtful whether the change will be rapid, and in any case we of the present day must face the difficulties of the transition state. Perfectly clear fundamental ideas on the part of the teacher are essential to success in this field of operations.

Multiplicity of Force-Units.

Moreover, the teacher should have a well thought out plan of campaign, though he should be able and willing to change this plan as occasion seems to require. It is a great mistake to insist that the pupil must get his ideas in the same order in which a master of the subject may choose to arrange his own matured conceptions. Such a master is apt to be too subtle and guarded in his preliminary statements, to look so far ahead as to raise difficulties which have not yet occurred to the pupil, — difficulties the too early consideration of which confuses and discourages the beginner. It is well to begin with simple and rather dogmatic statements, to be supported by experiment and argument and illustration as these are consciously or unconsciously demanded by the class. Simple problems, too, should be given in abundance, in order that the pupil may acquire that firm grasp of ideas which comes only by use.

Need of Simplicity.

To give a pedagogical syllabus of elementary ideas and relations in dynamics would be foreign to the purpose of this book; but the tabulation of a few important equations for each of several systems of units may serve a useful purpose, by showing similarities and differences, and even by exhibiting the complexity of the present situation in every-day dynamics. In the equations which are given below, acceleration, whenever mentioned, is assumed to be uniform acceleration, and force, whenever mentioned, is assumed to be uniform force.

Tabulation of Equations.

Moreover, the velocity, v, is supposed to be o at the beginning of the time, t.

Equations for Acceleration, Distance Travelled, Velocity Acquired, Force, Work, and Kinetic Energy, with the Absolute C. G. S. System of Units,

the dyne being the unit of force and the erg being the unit of work and of energy.

(1) a = acceleration $= \frac{v}{t}$, or $v = a\,t$.

(2) d = distance travelled $= \frac{v}{2} \times t = \frac{1}{2}\,a\,t^2$.

(3) $v^2 = 2\,a\,d$, from (1) and (2).

(4) $f = \frac{m\,v}{t} = m\,a$, where v is the velocity given to the mass m by the force f in the time t.

(5) w = work $= f d$.

When w is entirely spent in giving kinetic energy to m, we have

(6) $k.\,e.$ = kinetic energy $= w = f d = \frac{m\,v}{t} \times \frac{v}{2}\,t = \frac{1}{2}\,m v^2$.

Corresponding Equations with the Gravitation C. G. S. System of Units,

the gram-force, a force equal to the pull of gravitation on a gram mass, being the unit of force and the gram-centimeter being the unit of work and of energy.

(1) a = acceleration $= \frac{v}{t}$, or $v = a\,t$.

(2) d = distance travelled $= \frac{v}{2} \times t = \frac{1}{2}\,a\,t^2$.

(3) $v^2 = 2\,a\,d$, from (1) and (2).

(4) $f = \frac{m\,v}{g\,t} = \frac{m}{g}\,a$, where v is the velocity given to the mass m by the force f in the time t.

(5) $w =$ work $= fd$.

When w is entirely spent in giving kinetic energy to m, we have

(6) $k.\ e. =$ kinetic energy $= w = fd = \frac{mv}{gt} \times \frac{v}{2} t = \frac{mv^2}{2g}$.

With the Absolute Foot-Pound-Second System,

in which the poundal is the unit of force and the foot-poundal is the unit of work and of energy, we have precisely the same equations as with the absolute C. G. S. system.

With the Gravitation Foot-Pound-Second System,

in which the pound-force, a force equal to the pull of gravitation on a pound mass, is the unit force and the foot-pound is the unit of work and of energy, we have precisely the same equations as with the gravitation C. G. S. system.

Many engineers, in this country at least, keep to the gravitation English unit of force, and yet write

force $=$ *mass* $\times$ *acceleration.*

Mass in the Language of Engineering.

This is as if we should write equation (4) of a gravitation system in the form $f = \frac{m}{g} a$, and call $\frac{m}{g}$ the mass. That is, the engineer calls the mass of 10 pounds of iron $10 \div g$. It is to be hoped that in time there will be agreement between physicists and engineers as to the meaning of so important a term as mass.

CHAPTER XII

PLAN AND EQUIPMENT OF A LABORATORY

LET us suppose the school, for which we are to provide a laboratory, to be one of considerable size.

We have elsewhere, see Chapter VII., seen reason to believe that, for the best results, the laboratory sections should number not more than fifteen, though we may well make provision for slightly larger sections in view of emergencies.

Working Tables.

We will consider first the laboratory tables. Very short tables are comparatively expensive; very long ones are too much in the way when one has to go around them. The width should be such as to give plenty of room for a row of pupils on each side, with somewhat bulky apparatus before them, and without the necessity of crowding Bunsen burners and steam-boilers, for example, into close proximity with other articles which might suffer from the association. A good size for the table is 10 feet by 4 feet, the height being 3 feet. Such a table will give working room for six pupils, three on a side. Fig. 18 shows such a table in elevation and Fig. 19 shows it in plan. In Fig. 18, *gg* is a gas-pipe having six outlets downward for Bunsen burner connections, and four short horizontal branches (see also Fig. 19) for ordinary illuminating jets. In the same figure, 18, *bb* is a wooden bar, attached to the end posts by means of clamps, and adjustable at any height above the table between the gas-pipe *gg* and the tops of the posts. From this bar six brass rods project horizontally (see Fig. 21), each 1 foot long and each provided with a miniature vise, a thin saw-cut 1 inch deep, crossed by a pinching screw. This vise is not, perhaps, important, but the brass rods are very convenient for making suspensions, of spring-balances, for example, in careful weighing. The scale of Figs. 18 and 19 is 1 cm. for 1 foot. Few, if any, features of this table are original with me.

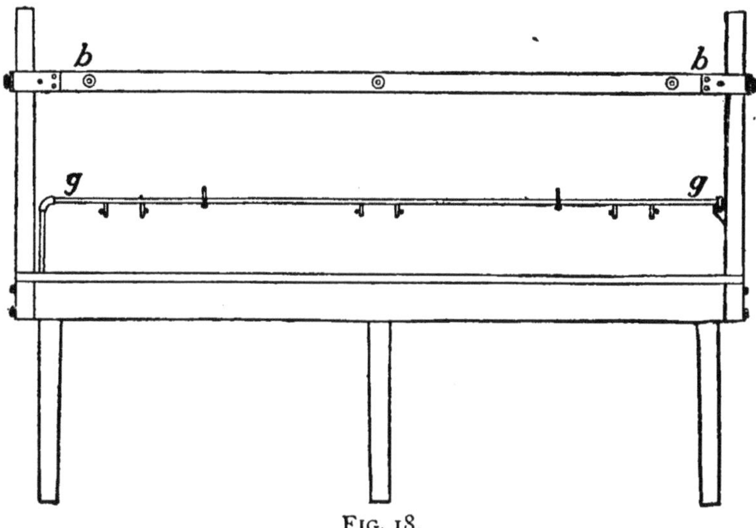

FIG. 18.

The four small circles shown in the table-top in Fig. 19 indicate holes bored through for suspension of pans bearing weights in a certain exercise on the bending of rods. The

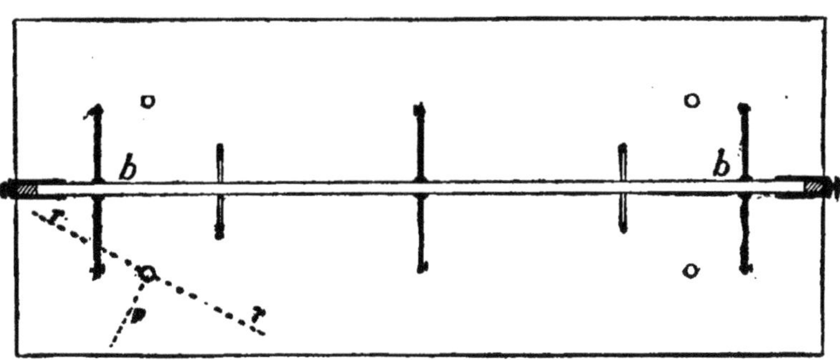

FIG. 19.

line *rr* indicates the position of a rod under observation, *p* being position of index. Two holes at mid-length of the table-top are not shown in Fig. 19.

The table should be so made as to keep a reasonably flat top, as in some exercises a level surface is very desirable, and therefore it should have as many as six legs. Pine and ash are good materials. Oak is objectionable on account of its tendency to warp.

If the plan of the course to be given involves furnishing each member of the class with a particular set of apparatus, which he alone is to use and for which he must be responsible, it may be necessary to provide drawers or lockers in or under the tables; but such a plan of work is, I believe, uncommon, and I greatly prefer plain tables with no such receptacles; for these latter interfere with certain uses of the tables and, being necessarily without glass fronts, hide whatever may be within them.

Let us suppose that we have three of these tables, accommodating, if need be, eighteen pupils in individual work.

The Laboratory Room.

If now we had a very long room lighted on one side, we might put all the tables in line near the windows; but this would not be a very good arrangement, for it would put one line of pupils with their backs to the light, and the other line with their faces to the light, — a disposition of the class unfavourable in some exercises to those facing the windows; for sometimes the parts of the apparatus demanding their most critical observation would be in the shadow of other parts. We will suppose the room (see A in Fig. 20) to be oblong, lighted on one side and one end, and will place the tables, 1, 2, 3, crosswise of this room, with one end of each distant about 3 feet from the lighted long side. A shelf supported by brackets on the wall is very useful, and we will suppose such a shelf, 15 inches wide, to run along the lighted end of the room at the height of the working tables. See 7. Our room should contain also a large soapstone sink, 5a, with an adjacent slop-table, 5, for holding battery materials, etc., a table for reference books, 4, another, 6, for demonstration apparatus, a wall blackboard, 9, and a long row of cases, 8, for storing apparatus. Ample provision of space for all these things, arranged as in Fig. 20, A, gives us a room 35 feet long and 25 feet wide.

The apparatus cases should be about 2 feet deep, easy range for the adult arm, and the top not more than 6.5 feet above the floor. Much bulky apparatus, of such a nature as not to be easily injured, can well be placed on top of the cases. The shelves should be adjustable at various heights, unless some one knows the apparatus well enough to place them in advance. The highest shelf should not be more than 5 feet above the floor.

Apparatus Cases, etc.

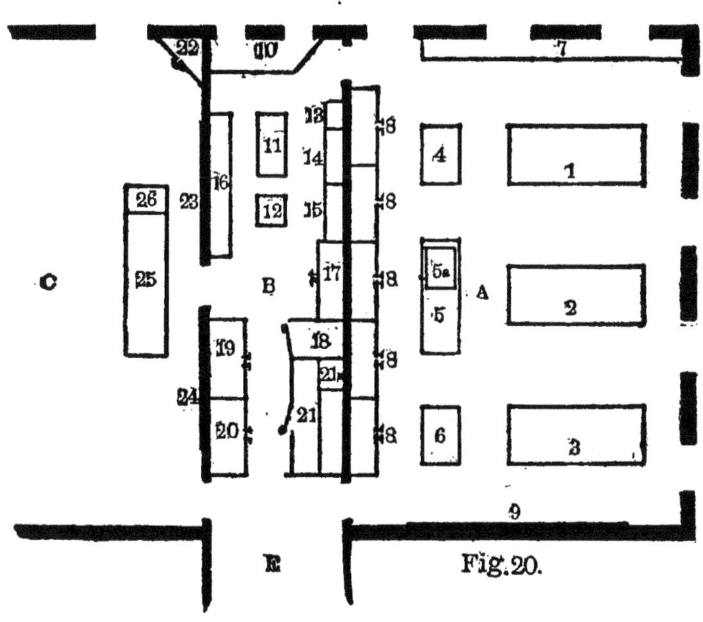

Fig. 20.

A few drawers, for cork-stoppers, rubber tubing, small hardware, etc., and a few cupboards for glassware, crockery, record books, and other things more useful than sightly, can be placed in, or under, the tables for books and demonstration apparatus.

Nothing has yet been said here in regard to the height of the laboratory ceiling, or the number and dimensions of the windows. It goes without saying that the room should be well lighted. Fig. 20 indicates seven windows, each 4 feet wide. As to the height of the ceiling, there are few experiments which demand greater height than that of the ordinary room in a modern, well constructed, school building, and it would be hardly justifiable to make an exceptional height for the sake of these few experiments.

We must presently consider the lecture-room and the preparation-room, or workshop. As the latter is a necessary adjunct to both the laboratory and the lecture-room, it may well be placed between them, if this arrangement is consistent with the general plan of the school-building, of which we assume the rooms for physics to be a part. A common form for such a building, in the case of public schools, is a long main body, with rather broad hall-ways, or passageways, running along its rear, and with a wing at each end. At the end of a passageway (see E, Fig. 20), and in line with it, there is likely to be a long narrow space, sometimes utilized as a coat-room. This space, which I shall assume to be 10 feet wide, will here be taken as a workshop and general utility room (B, Fig. 20). Circumstances must determine whether the lecture-room or the laboratory shall occupy the rear of wing.

Workshop.

The plan shown in Fig. 20 does not undertake to provide for instruction in the use of tools, or for the manufacture of much apparatus, but only for such operations of construction and repair as the energetic teacher must be prepared to undertake. This equipment should include a work-bench and a lathe, with tools for working in both metal and wood, an emery wheel for sharpening tools, and facilities for soldering and glass-blowing. The last two operations may use a blast lamp in common, and should therefore be carried on near each other. The blast of air for the lamp can be furnished by means of a Richards pump, with compression chamber, placed in a sink. From such a pump, with a good head of water, a sufficient current of air for the lamp can be carried many feet through a half-inch pipe. The teacher should have also at his service an outfit for photographing and for blue-printing, and the dark room for developing may well be placed at the inner end of the work-room. Of course there should be shelves and a case of drawers for stock and tools.

All these things being placed, and also an electric motor for power, which may be on the floor or on a platform above the rest of the machinery, there will remain in room B a considerable amount of space available for storing the lecture-room apparatus. The most convenient way of getting such apparatus into the lecture-room, C, is through a door in the middle of the wall behind the lecture-table.

The width of the lecture-room, as shown in Fig. 20, is 35 feet. The depth I leave undetermined, as that should depend on the number of pupils it will be required to hold. The common practice of making the width of a lecture-room much greater than its depth is unfortunate. The front seats at the sides in such a room are undesirable, for they give a very oblique and therefore indistinct view of things on the blackboard which is behind the lecture-table.

Lecture-Room.

The lantern-screen is also supposed to be behind this table, on a roll which draws it up out of sight when it is not needed; but, if the ceiling is high, the wall space above the blackboard, if finished smooth and white, serves exceedingly well as a screen for projections. The lantern itself is supposed to be at the rear of the room. The one window shown in the wall of C is supposed to be on a level with the lecture-table, so that a mirror placed at this window will put the sun's rays at the service of the lecturer, provided this window has a southerly outlook.

At the lecture-table sink it is well to have two faucets, one of which should end in a screw, so as to admit of ready connection with an aspirator. It is well to provide the table with a horizontal adjustable bar carried by end posts, like the bar and posts of the working-tables (see Figs. 18 and 19); but as these objects might sometimes obstruct the view of something on exhibition beyond them, the posts should be so attached to the table as to be easily removed or replaced.

Some Equipments.

For a source of low voltage electricity, which one needs to use occasionally at the lecture-table, the arrangement illustrated by the following diagram (Fig. 21) is to be recommended: S_1 and S_2 are two storage cells of the ordinary type, connected with each other in series through the binding-posts b_2 and b_3. These cells are charged by a battery, DD, of six large gravity Daniell cells connected with each other in series. The connections here shown are maintained all the time, so that the storage cells are always ready for use. If two volts are needed for use in any experiment, connection is made with the binding-posts b_1 and b_2, or with b_3 and b_4. If four volts are needed, connection is made with b_1 and b_4. All the cells should be covered so as to diminish evaporation.

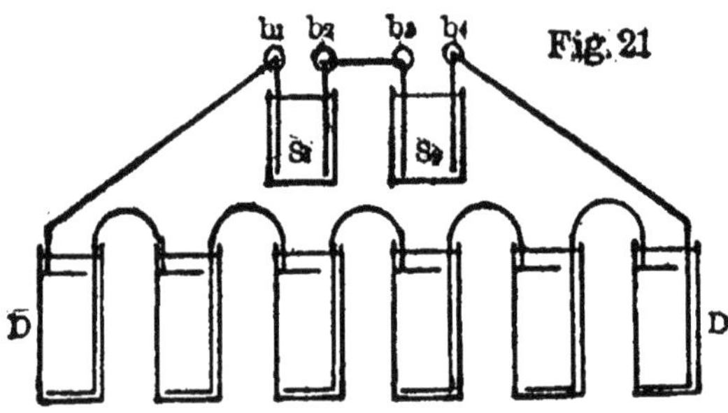

Fig. 21

A convenient and useful device for controlling strong currents by variation of resistance is in the form of a column of carbon plates, each about four inches long and ¼ inch thick, held in a frame between two thick end plates of brass, one of which can be pressed against the adjacent carbon plate by means of a screw passing through one end of the frame. The greatest defect of this device is that one cannot tell, without some supplementary measuring instrument, how great its resistance is at any given instant.

The reflecting galvanometer (see Figs. 15 and 16, Chapter VIII.), if the lecture-room is provided with such an instrument, can be placed, when in use, on a shelf just below the blackboard, the inclined screen being placed above the blackboard, or the whole outfit can be put in one corner of the room in place of the triangular apparatus case, 22 of Fig. 20.

If the school is one in which the physics class is small, the laboratory room, with little or no change from its plan as already given, can be used as a lecture-room.

I have said little in this book in regard to the general body of apparatus with which the physics department of a well equipped school should be provided, and I shall not now undertake to treat of this matter at length. **Apparatus.** Accompanying the Harvard *Descriptive List* of laboratory exercises is a detailed list of apparatus for these exercises, which is the product of years of experience and suggestion and gradual development. Much of this apparatus, which is simple and inexpensive, would be useful in any beginner's course of laboratory physics, looking toward college or not looking toward college. Moreover, several manufacturers of school apparatus keep the articles of the Harvard list in stock and know these articles by the numbers they bear in that list, a fact which facilitates ordering from them if one has the Harvard list at hand.

As to the apparatus for lecture-room purposes, almost any modern descriptive text-book is a fairly good guide, though of course one should compare various books and various catalogues of dealers in apparatus before making any large purchase.

CHAPTER XIII

PHYSICS TEACHING IN OTHER COUNTRIES

REFERENCES.

Board of Education of the English Government. Special Reports on Educational Subjects.

British Association Reports, in many places; for example, in 1889 and 1890, "Suggestions for a Course of Elementary Instruction in Physical Science."

Delalain Frères, 115 Boulevard Saint-Germain, Paris, official programmes of primary and secondary instruction in France.

Russell, J. E. German Higher Schools. New York and London, Longmans, Green & Co. 1899. Pp. 467.

Sharpless, I. English Education. New York, Appleton & Co. London, E. Arnold. 1892. Pp. 193.

WITHOUT undertaking an examination of the state of elementary physics teaching in all European countries, we may well inquire what it is in Germany, England, and France; for these are the countries to which, rightly or wrongly, Americans are in the habit of looking for suggestion and instruction.

Germany has a well established system of physics teaching in her schools as well as in her universities, and this system, if it is not the cause of her eminence in physical research and her success in commercial scientific undertakings, is at least contemporaneous with these achievements. England has for a number of years studied German methods of secondary education in science, hoping thus to find and profit by the secret of her dangerous commercial activity. It is well, therefore, that we should look first at these same methods as they are found in German schools below the universities. The numerous quotations given below, from Russell's *German Higher Schools*,[1] make this task easy.

It should be said in advance that in both classes of the higher schools, the classical gymnasiums and the *real*-schools,[2] the full course of study is nine years, which, beginning with the year of the lower class, are numbered thus, *sexta*, *quinta*, *quarta*, *unter-tertia*, *ober-tertia*, *unter-secunda*, *ober-secunda*, *unter-prima*, *ober-prima*. Many pupils, however, leave the schools at the end of six years, there being a well defined break in the course at that point.

(From p. 330.) "The chief aim of all instruction in the natural sciences [including physics] is to cultivate the habit of keen and accurate observation, to strengthen the pupil's reasoning powers and to increase his ability of expressing clearly what he sees and thinks. The acquisition of a fund of systematic knowledge or useful information is a secondary consideration."

After remarking (p. 333) that "the science work in the Real-schools is taken more seriously than in the Gymnasien," the author gives, "as a type of what is done in Prussia the course of study prescribed in the Königstädtisches Realgymnasium of Berlin." The physics of this course is,

"Unter-secunda. [Sixth year]. *Physics*, 3 hours [per week]. First semester: Frictional electricity and phenomena out of [taken from] the domain of magnetism and galvanic electricity. Acoustics and optics. Second semester: Mechanics of solid, liquid, and gaseous bodies. General properties of matter. Parallelogram of forces and of motion. Laws of falling and vertically projected bodies. The simple machines. Text-book, Jochmann, *Grundriss der Experimental Physik*."

"Ober-secunda. *Physics*, 3 hours. First semester: Magnetism and galvanic electricity. Second semester: Heat, repetition and extension of mechanics, especially of oblique projection and of central motion. Text-book, same as in *Unter-secunda*."

---

[1] Longmans, 1899.

[2] The *Realgymnasien* and the *Oberrealschulen*.

"Unter-prima. *Physics*, 3 hours. First semester: Wave theory, acoustics and optics. Second semester: Mechanics. In both semesters, reviews and more thorough mathematical treatment of particular parts of the earlier work. Solution of problems. Text-book, same as in *Unter-secunda.* (Physical laboratory exercises, 2 hours, optional.)"

"Ober-prima. *Physics*, 3 hours. First semester: Optics. Second semester: Mechanics. In both semesters, reviews and more thorough discussion of parts of the earlier work, especially quantitative determinations and methods of measurement. Text-book, same as above. (Physical laboratory exercises, 2 hours, optional.)"

(From p. 345.) "According to the Prussian syllabus of 1892, the course in physics is divided into two parts. The first part is intended to give the pupil some notion of the fundamental principles of the subject as exemplified in the ordinary and more familiar manifestations of nature; it is concluded with *Unter-secunda.* The continuation of the course aims to give those who may pass on to the university a more comprehensive understanding of physical laws and their applications. This division is in strict accord with a prevailing idea of the Berlin Conference [in which the present Emperor figured so prominently], that those leaving school at sixteen should have as symmetrical a training as it is possible to provide. Only the most important principles are taught in the first part of the course, and much stress is put upon the application of these to the practical affairs of every-day life.[1]

"The advanced course is first of all a repetition and extension of the earlier work, and in the second place a more extended mathematical treatment of the subject. This latter phase of the work can be done successfully only in the *Real*-schools, inasmuch as the mathematics taught in most *Gymnasien* is insufficient for the purpose."

---

1 "Full information of what may be accomplished in this preliminary course may be found in the *Zeitschrift für den physikalischen und chemischen Unterricht, Jahrgang* V, Heft 4 (April, 1892)."

(From p. 346.) "A text-book is always employed in teaching physics and chemistry, precisely in the same manner as in teaching natural history. But, unlike the methods commonly used in American and English schools, German teachers invariably use these books for reference only. It is not expected, however, that they will take the place of the elaborate compendiums found in each school-room; they are mere outlines of the subject, intended to assist the pupil in making scientific classifications, not for purposes of recitation. In fact, as we have repeatedly observed, the German teacher never assigns a lesson in advance to be studied at home. Recitations, therefore, at least in the American sense, are unknown.

"A typical lesson always includes a review of the principles and experiments of past lessons which have a direct bearing upon what is next to be presented. The teacher explains the nature of the apparatus with which he is to deal, and places it upon his desk in full view of the entire class. . . . Certain conditions are stated, and the class questioned as to what results may reasonably be expected. This preliminary discussion having carefully prepared the way for a right understanding of the experiment, the demonstration by the teacher follows. The students are required to make note of the apparatus used, the principles involved, the conditions under which the reaction occurred and the results obtained. By means of a running fire of questions, the teacher keeps himself informed in regard to the mental state of his class; for it is his duty to see not only that all understand the trend of the experiment, but also that its significance is realized.

"German practice is always consistent in its adherence to the idea that good teaching never leaves the pupil in doubt. In mathematics he is not assigned a problem to wrestle with by himself alone," etc.

"Every principle worth demonstrating is illustrated in class. But the teacher does more than demonstrate; he *teaches* as well. And successful teaching requires that present impressions be definitely related to past experiences. Wrong relationships, or none at all, are an inevitable consequence of misapprehension. For this reason the German teacher counts it his duty to prevent his students drawing wrong inferences. They have not yet arrived at the stage of independent study; that comes in the university. In secondary schools no time should be wasted in beating about the bush. The ability to make an occasional lucky guess is in nowise identical with sustained logical thought.

"At the conclusion of a lesson topic, the pupil is directed to consult his text-book and afterward write up his notes. This done, the teacher inspects the book at his leisure.

"Laboratory exercises, if required at all, are introduced at this point, in order that students may themselves duplicate the experiment performed by the teacher or make other demonstrations putting to practical test the knowledge just acquired. The function of laboratory practice, as will be seen, is to make application of facts already learned, not at all for the purpose of presenting new truths or arriving at new deductions. Inasmuch as laboratory practice is optional, and the exigencies of the time-card usually place it out of school hours, few students enter for it."

(From p. 348.) "Probably the best adducible evidence of the relative value of the various studies, as popularly estimated, is the part each plays in the final examination. Judged in this way, the sciences take a low rank. Physics may be counted as a fourth part of mathematics in the gymnasial examination; in the *Real*-schools, one problem is assigned in physics and one in chemistry.[1] The worst of it is that, 'nothing short of a miracle,' to quote a German teacher, 'can prevent the promotion of the most deficient member of the class, provided his attainments be satisfactory in other subjects.'"

There is much to be commended in the physics instruction which the German boy receives. It does not attitudinize, does

[1] A foot-note here gives problems set at the final examination (Abiturientenprüfung) of a *Realgymnasium*. The problems in physics are simple, involving the use of Ohm's law, the tangent galvanometer and Wheatstone bridge.

not call itself by a name which it cannot live up to; it drives straight and hard at some of the most important objects of study, a useful knowledge of physics and a useful habit of looking at and thinking about those physical phenomena which are presented to the pupil's view. I do not feel disposed to criticise German school-teaching as too little "inductive" or "heuristic," though possibly it may be so. Its chief defect, and a serious one, seems to be that it does not give the pupil laboratory work for his own hands, and therefore leaves him wanting in that actual experience of apparatus which is so important for any one who must conduct or devise experiments or make any objective use of physics. Transported to America, where the incentives to scholarly effort on the part of young pupils are at present much less strong, and where teachers are less thoroughly equipped, than in Germany, the German school system of physics teaching would probably not work well.

When we turn from Germany to England, and attempt to realize the state of science teaching in the schools of the latter country, the field of view grows suddenly obscure. For, as compared with Germany, England can hardly be said to have a system of education: she has rather a state of development, and in some respects a rapidly changing state, the changes being as rapid in science instruction as in any other.

The *Special Reports on Educational Subjects* issued by the English Government[1] contain a good deal of interesting matter relating to instruction in science. In volume 6 (1900) Mr. Archer Vassall, Assistant Master at Harrow, writes as follows:

"In Public Schools [the endowed schools like Rugby, Eton and Harrow] the teaching of science has only recently begun to take reasonable shape, and ceased to be a series of fireworks, or isolated physical phenomena, presented in a casual and indigestible manner to the pupil; while there has been so little of it in the Preparatory Schools [preparatory to

[1] Vols. 1–3 by the "Educational Department," later volumes by the "Board of Education."

the endowed Public Schools] that its past and present state in these institutions does not require any long exposition.

"Nevertheless, now that the large number of subjects included under the head of Science are more reasonably taught to elder boys and others, there has arisen a fairly widespread feeling, amongst both parents and schoolmasters, that some elementary information on scientific subjects should be given to boys whilst still at Preparatory Schools, and that these subjects afford valuable material for educating the minds of such boys. To their credit be it said, Board Schools and Girls' Schools have for some time realized this fact, and in many of them scientific subjects find a place in the curriculum.

"In Preparatory Schools the result of this inclination has been that several tentative efforts in scientific instruction have been made, and are still in progress at many of them, though nothing approaching the systematic 'nature study' of the young American has as yet been achieved."

Mr. Vassall's opinion, as shown in this article, is in favour of physics, for preparatory schools, rather than chemistry, and strongly for laboratory work combined with lectures, rather than lectures alone.

Volume 2 of the *Reports* contains (pp. 389–413) an article on "The Heuristic Method of Teaching or The Art of Making Children Discover Things for Themselves," by Professor Armstrong. The "British Association Scheme" of science instruction, to which he frequently refers, was the outcome of the work of a committee of the association, which was appointed in 1887 and reported in each of the three following years. In 1889 and 1890 the committee printed in its reports *Suggestions for a Course of Elementary Instruction in Physical Science*, by Professor Armstrong, who was a member of the committee. It is evident that these reports have had much influence in England, and Professor Armstrong, in the article named above, claims a very marked degree of success for teaching inspired by the methods and principles set forth in "The British Association Scheme." He states in this paper that in 1897 this scheme "was in operation in no fewer than 40 of the London Board Schools."

In general terms this scheme, as originally set forth, is to train the pupil from childhood to observe, think about, and experiment on, common things, air and common liquids and earthy materials, for example, not with a view to making him by and by a specialist in chemistry or physics, but for the purpose of forming certain important habits and cultivating certain important powers, while giving a considerable amount of directly useful information. It is admitted that progress will be slow, as it is in all the other important studies of childhood, but great things in the way of preparation for the inevitable and unending conflict of nations, in commerce, industry, and war, are hoped for by the advocates of this scheme of instruction, if it is undertaken and persistently carried out.

The title of Professor Armstrong's paper should be read in the light of the following passage (p. 407), by which it appears that the "method of discovery" by individual pupils is not rigidly adhered to. "No books will be used, but the class will gradually write its own book and so come to understand how books are written; for whenever an object has been properly studied, the teacher, instead of dealing with the scholars individually, will call them to order as a class, and by judicious questioning will then elicit all that is needed for the description of the work done. The simplest possible account will be written on the blackboard as the questioning proceeds, and at the close of the lesson a senior pupil will copy this with a typewriter, and each member of the class will afterwards receive a copy, which will at once be pasted in a book, to be kept for reference and used as a reader."

Appendix A to the paper of Professor Armstrong gives the "Course of Instruction in Elementary Science adopted [in 1896] by the Incorporated Association of Headmasters" of secondary schools[1] for pupils commencing the study of physics.

[1] According to Sharpless, *English Education* (Appleton, 1892), "Secondary education [in England] is now in the hands of a number of

Professor Armstrong speaks of this syllabus as "based on the British Association scheme." It is worthy of careful examination. The headings under Elementary Physics are:

1. Measurement of Length.
2. Measurement of Area.
3. Measurement of Volume.
4. Measurement of Mass.
5. Measurement of Density.
6. Measurement of Thrust and of Pressure, of Pull and of Tension. Distinction between solids, liquids, and gases.
7. Measurement of the force which a liquid exerts upon a body immersed in it.
8. Measurement of Temperature.
9. Measurement of Quantity of Heat.
10. Measurement of Vapour Pressure.
11. Measurement of Force in pounds or grams weight, and their Graphic Representation.
12. Resolution of Forces.
13. Equilibrium of Three Forces.
14. Equilibrium of Four or more Forces.
15. Parallel Forces.
16. Centre of Gravity.
17. Principle of Moments, Levers.
18. Simple Machines.

According to Professor Armstrong (p. 397) the Oxford and Cambridge Local Examination Authorities have tried to make examinations suited to the science syllabus of the Headmasters' Association. "Unfortunately, however, their instruction in no way whatsoever imply or involve heuristic teaching; and it is only too clear that that which is fundamental in the recommendations of the British Association scheme has not been understood." This was printed in 1897–1898; things may be different now.

---

private schools of all degrees of goodness and badness, of a few nonconformist denominational schools which are usually good, and of the endowed schools for boys" (Public Schools and Grammar Schools).

The general tendency of Professor Armstrong's writing seems to be to discourage the use of printed books and to make the pupil distrust accounts of what has not been seen by himself or his classmates. American teachers should, I think, be slow to follow this suggestion. The ordinary American boy is only too willing to act on the hint not to study books on physics. He will do laboratory work cheerfully enough, even when he has only the dimmest idea what it is all about, but he shrinks from the, to him, painful effort of getting from a book the definitions and the reasoning necessary to make the laboratory work intelligible. This is not because he has any predilection for the "heuristic" method; for he delights to be told things by word of mouth instead of seeking them out for himself, and, if not persistently discouraged from the practice, will habitually stand with an open book before him and ask for information that is plainly given on the printed pages beneath his eyes. Nor does the disinclination to reading necessarily disappear with youth. It often persists into manhood and renders fruitless the labour of years.

Is it not quite possible that the scientific pre-eminence of the Germans as a race is due largely to their habit of reading widely and thoroughly, of mastering by reading not only the bulky treatises and periodicals of their own language, but also the scientific publications of foreign tongues? Consider the significance and influence of such a journal as the BEIBLÄTTER to the ANNALEN DER PHYSIK. What testimony it gives to the zeal of the Germans in the study of science through the printed page.

The completeness with which the educational system of France has been worked out by a thoughtful and ingenious people makes it worthy of study. The official programmes of primary and secondary instruction, with which we are concerned, are set forth in considerable detail in frequent publications issued by Delalain Frères, 115 Boulevard Saint-Germain, Paris. Any one can by reading these programmes get a fair idea of what is being done in schools of any given class throughout France.

In examining with considerable care the parts relating to physics in the official programmes of the various kinds of schools, I have found nothing anywhere which seems to require or to provide for experimental work to be done by pupils, though I am informed that laboratories for pupils in physics do exist in some *lycées*. Indeed, for students taking the classical course I find nothing under chemistry, even, which seems to make provision for laboratory work by pupils until the *Classe de Mathématiques Speciales*, for young men who have completed the ordinary *lycée* course, is reached. Here twelve chemical "manipulations," each, apparently, occupying the student four hours, are strictly prescribed.[1] In the "modern" course of the *lycées* there is some little provision for laboratory work in chemistry in the third class, the second class, and the first class (*sciences*), about eighty hours in all.

In the elementary primary schools there are object lessons (*leçons de choses*). Although the object of the primary school instruction is frankly and emphatically utilitarian, an attempt is made to cultivate the philosophical imagination of the pupils. "In all instruction, the master, at the beginning, makes use of tangible objects, has the children see and touch the things, puts them in the presence of concrete realities; afterward, little by little, he exercises them in getting at the abstract idea from the

[1] An official order relative to these exercises, which order was written in 1854, and is apparently still in force, runs thus: "The pupils ought never to be left to themselves during the manipulations. These should always be preceded by a conference, in which are set forth, with all necessary details, the operative processes relative to the manipulations which the pupils are to perform. In describing these operations the professor executes them, making use of the same apparatus which the pupils are to use. Finally, the apparatus, mounted in advance, is displayed before their eyes, which indicates all the dispositions which they will have to observe in the arrangement of the pieces which compose it." This order affords a good example of the care and precision with which official instructions are issued to teachers in the public schools in France.

objects, in comparing and generalizing, in reasoning without the aid of material examples."

In the higher primary schools for boys we find

"Physics and Chemistry. (Two hours a week [for both, not for each] during the three years.) *General Remark.*— In each year the course in physics and chemistry will be essentially experimental." That is, the lessons are to be illustrated by lecture-table experiments.

"Physics. FIRST YEAR. *Heat.*— In general, bodies expand under the influence of heat.— Simple experiments.— They expand unequally.

"*Temperature.*— Mercury thermometer. — Graduation. — Centigrade Scale, degree centigrade.— Maximum and minimum thermometers," etc.

The general course is the same for all three sections, commercial, industrial, and agricultural, but with applications varying from one section to another, "according to the special needs of each."

In the higher primary schools for girls the physics of the first two years is identical, so far as the official programmes of topics show, with that of the same years in the corresponding schools for boys, but the like is not true of the third year.

As the time allowed for physics and chemistry together in these schools for girls is only one hour a week for the three years, just one-half as much as the time given to the same subjects in the corresponding school for boys, it is evident that the instruction received by girls is comparatively superficial.

In the first three years, "*Division Élémentaire*," of the course in the *lycées* and *collèges* for boys, there are object lessons, one hour a week, which include a little physics. A note of instruction given in connection with the science of the preparatory class, and repeated over and over again with reference to the science of subsequent classes, up to the *Classe de Rhétorique*, is the following: "Professors are especially charged to spare no pains to make the demonstrations and the relations of facts well understood, and *not to dictate their courses.* They may, if they think it best, put into the hands of the pupils an autographic text or a book which will relieve them from the necessity of developing personally all parts of the course."

During the next three years, *Division de Grammaire*, there is in the classical course, and also in the "modern" course, a little of zoölogy, of botany, and of geology, but there is nothing of physics, as such, or of chemistry, in either course.

Indeed, there is in the classical course of the *lycées* no study of physics under its own name until the *Classe de Rhétorique*, the ninth year of the course, is passed; then the student has, in the *Classe de Philosophie*, five hours a week divided between physics and chemistry, or, in the *Classe de Mathématiques Élémentaires*, in addition to some study of elementary mechanics, six hours a week divided between physics and chemistry. Of course the physics work done in these classes is elementary, a fact sufficiently illustrated by the following list of "*Compléments*," with which the physics programme of the *Classe de Mathématiques Élémentaires* ends:

"Laws of falling bodies. — Atwood's machine. — Morin's machine.

"Proportionality of forces to accelerations. — Mass. — Its measure by means of weight.

"Pendulum. — Applications.

"Very elementary notions of work, vis viva, energy, the mechanical equivalent of heat.

"Various forms of energy. — Principles of the conservation of energy.

"The steam-engine. — Condenser. — Expansion."

This brings the student in the classical course to his baccalaureate. His opportunities to learn physics have been, apparently, about the same as those afforded by American colleges having the old-fashioned non-elective course.

In the "modern course" of the *lycées*, physics, as such, is taken up earlier, three hours a week being divided between it and chemistry in the *Classe de Troisième*, the seventh year. The spirit in which it is to be taught is indicated by the following official extract from some report : "The professor of sciences will not lose sight of the fact that the object of his instruction is not solely to teach his pupils a certain number of acquired facts, but that it is also, particularly in the course of modern studies, where the sciences hold a large place, to contribute to the general culture of the mind. He will, therefore, so act that the high educative virtue peculiar to science, which those profit by who give themselves up to it, shall be in force as much as possible in his teaching." This precept is meant to apply to the science teaching in general; but under the head of physics and chemistry the following direction is given: "To the demonstration of scientific truths the professor will add upon occasion the exposition of methods [of measurement or research] and the history of discoveries." The physics of this third class consists of the elementary study of gravity and heat. In the next year, second class, when four hours a week are divided between physics and chemistry, the work deals with electricity, magnetism, acoustics, and optics.

On leaving the second class (eighth year) the student may enter the first class (letters) in which there is no physics, the first class (sciences), in which four hours a week are divided between physics and chemistry, or the *Classe de Mathématiques Élémentaires*, in which, as we have already seen, six hours a week are divided between physics and chemistry. So far as one can see by the official programme, the physics course of this last class is precisely the same for those who have come through the modern course as for those who have come through the classical course, although it has been preceded by considerable physics study in the modern course, and by almost none in the classical course.

In the *lycées* and *collèges* for girls there is no physics, as such, until the third year, when the pupils are about fourteen years of age. In this year two hours a week are divided between physics and chemistry, the former science, apparently, having the greater part of the time. The subjects taken up in physics are all under the headings *Gravity* and *Heat.*

In the next year, the fourth of the course, one and a half hour per week is given to physics and chemistry, the physics topics treated being all under the headings *Acoustics* and *Optics.* In the next year, the last year of the regular course, physics and chemistry together have two hours per week, the physics relating to magnetism and electricity.

The total time, then, for physics and chemistry in the *lycée* course for girls amount to five and a half hours a week for one year. This is about one-quarter of one year's work, the total number of hours of stated instruction being twenty-one per week in the third year and twenty-four per week in each of the two following years.

It appears, then, that the physics work of French schools is light and expansive, descriptive, somewhat historical, somewhat philosophical. It is, no doubt, skilfully conducted. Yet we need not be surprised that philosophers[1] have little respect for it, as a means of sound discipline, in comparison with Latin and Greek; for it is evidently intended to be an entertaining and informing rather than a formative study.

On the whole, it appears that the best secondary schools in America, in trying the experiment of teaching physics by means, in part, of laboratory work done by the pupils, have little or nothing to learn from the corresponding schools in France, Germany, or England. For France has apparently never dreamed of such an undertaking, Germany has never seri-

[1] See Fouillee, *Education from a National Point of View,* Chapter II.

ously considered it, and England is no farther along with it than we are in America, if indeed she is as far. If we make a final and permanent success of this venture, as we seem likely to do, Europe will have an opportunity to learn from us, and we may in this way be able to repay in some small measure the educational debt which we have owed to her so long.

Index

[CHEMISTRY, pp. 1–227. PHYSICS, pp. 229–371.]

www.ingramcontent.com/pod-product-compliance
Ingram Content Group UK Ltd.
Pitfield, Milton Keynes, MK11 3LW, UK
UKHW021522300726
14060UKWH00012B/598